Springer Series in Optical Sciences   Volume 22

Edited by David L. MacAdam

Springer Series in Optical Sciences

Edited by David L. MacAdam

Editorial Board: J. M. Enoch   D. L. MacAdam   A. L. Schawlow   T. Tamir

Lasers in
Photomedicine
and
Photobiology

Proceedings of the European Physical Society,
Quantum Electronics Division, Conference,
Florence, Italy, September 3–6, 1979

Editors
R. Pratesi and C. A. Sacchi

With 108 Figures

Springer-Verlag Berlin Heidelberg GmbH 1980

Professor Dr. RICCARDO PRATESI

Istituto di Fisica Superiore dell'Università
e Laboratorio di Elettronica Quantistica,
Firenze, Italia

Professor Dr. CARLO ALBERTO SACCHI

Centro di Elettronica Quantistica, Politecnico,
Milano, Italia

ISBN 978-3-662-13499-3   ISBN 978-3-540-38270-6 (eBook)
DOI 10.1007/978-3-540-38270-6

Originally published by Springer-Verlag Berlin Heidelberg New York in 1980.
Softcover reprint of the hardcover 1st edition 1980

Offset printing: Beltz Offsetdruck, Hemsbach/Bergstr.
2153/3130-543210

To
Giuliano Toraldo di Francia

Obituary for Sergio Pereira Da Silva Porto

We wish to remember in these pages our dear friend Sergio Porto, who died prematurely only a month before this Conference.

Porto was one of the most outstanding scientists in the field of Laser Spectroscopy. He was a pioneer in the use of laser techniques in Raman Spectroscopy and today Laser Raman Spectroscopy has become a powerful tool for current research in the structure of matter, in particular of biological substances.

In 1974, after a 20 year absence, he returned to his native country, Brazil, to establish a quantum electronics group at the University of Campinas. Among the various research fields he promoted with his well-known enthusiasm was the application of lasers in biomedicine to obtain new and "unique" (as he said smiling) results. We met him in Meriden in 1978 with his medical collaborators (Escudero and Freitas), where they reported the first case of successful tympanoplasty with argon laser in man[1]. One of us (R.P.) met him again at a Laser Spectroscopy Conference near Munich a few weeks before his death. He was happy to anticipate finding with Freitas a new method for argon laser therapy of glaucoma with a successful two-year follow-up. He was very interested in new applications of lasers in Medicine and Biology. His death is a great loss for the scientific community: he will live in the work of his collaborators in Campinas and in the memory of his friends everywhere in the world.

R. Pratesi   C.A. Sacchi

---

[1]L.H. Escudero, S. Porto et al.: Argon Laser in Human Tympanoplasty. Arch. Otolaryngol. *105*, 252 (1979)

Preface

This Conference on biomedical applications of lasers was organized by the
Quantum Electronics Divisional Board of the European Physical Society
(E.P.S.) and held at the Villa of Poggio Imperiale in Florence, September
3-6, 1979.

As known, laser surgery (especially microsurgery and endoscopic photocoagulation) has recently made important progress, and the field is expand-
ing rapidly.

Very significant applications of lasers have also been achieved in Biology during recent years (cell microsurgery, cell counting and sorting,
cytofluorimeter devices, etc.) and the potential of laser techniques in
this field is now sufficiently well established.

A new class of applications of laser radiation in Medicine has recently
been made possible by important results obtained with low intensity (non-
coagulative) visible lasers, such as photodynamic therapy of tumors. At the
same time important branches of Medicine, where light effects are studied
and optical techniques are presently used for a certain number of clinical
applications, such as dermatology and pediatry, appear to be still in their
infancy as far as the proper use of optical radiation and techniques, and
the understanding of fundamental photoinduced biological processes are concerned. Moreover, laser photobiology appears a very promising field for the
investigation of fundamental processes at the biomolecular level.

For these reasons the Conference topics have been chosen in order to
offer a discussion of the potential non-surgical use of VIS and UV lasers
in Medicine, and on the study of photobiological processes which can be induced by laser light, or investigated with laser techniques in Photobiology.
Open problems in most important biomedical branches where light effects
are studied and optical techniques are currently used for therapy are reviewed in this book, and the possible introduction of laser techniques is
critically discussed.

To complete the spectrum of laser applications and to attract more researchers and practicing physicians, an overlapping Conference on Lasers in
Bio-Medicine and Surgery was organized on September 6-7 under the sponsor-
ship of E.P.S. and the Special Project on Laser Medical Applications of the
Italian C.N.R.

The two joint Conferences helped to close the gap between the fields of
basic research and practical therapy, improving communications between researchers and physicians. The former, in fact, are often unaware of the
problems of treating patients, while the latter have often only a superficial understanding of the biological processes involved. More than 200
participants, representing 20 countries, attended the Conferences, and about
100 papers were submitted in total.

Here, only papers concerning the first Conference are presented; the
others which belong to already well established fields of research and applications are available in specific journals and Conference proceedings.

Since the meeting was organized by the Q.E. Division of E.P.S. and addressed to a large audience of laser physicists, several papers have been included to introduce the basic concepts in the field of Photomedicine and Photobiology. No introduction to the "laser" has been given: we assume its main properties, as a source of high-intensity and collimated light, very monochromatic under stabilized continuous wave operation, or capable of generating ultrashort pulses when operated at picosecond pulse regime, are well known. People interested in a review paper on Laser Sources may refer to the reference given in the footnote.

Invited speakers were chosen by the Conference organizers from among the most outstanding authorities and active researchers in the various research fields. Contributed papers were open to all, and reviewed in advance only to a limited degree, and more carefully before publication. Several papers, however, gave rise to interesting and sometimes controversial discussions. This is typical in interdisciplinary meetings where people apply well-known techniques and methods out of their own research field. We believe it is worth publishing even those papers which did not eliminate the source of controversy in the final edition, as we think they may stimulate further research and better understanding of the problems involved. Therefore, the reader must exercise judgement as to correctness of statements not independently confirmed. In particular, the concept of laser selectivity in photobiology turns out to need a more careful definition before being accepted by classical photobiologists.

Numerous people at the Quantum Electronics Laboratory of CNR (Florence) contributed to the organization; in addition to Umberto Vanni and Silvia Balloni, who acted as local organizers, we thank Miss Joanne Askham from the Physics Department, Heriot-Watt University, Edinburgh, for her help during the Conference. Grateful thanks to Prof. Desmond Smith and to Mrs. Gillian Smith for their kind advice and assistance in running the Conference.

We would like to thank all members of the Program and Advisory Committee who gave us their valuable advice in the preparation of the Conference Program.

We tried to provide the Conference members with one of the most beautiful settings available in Florence. In this regard we are grateful to the Sovrintendenza alle Belle Arti for putting at our disposal the magnificent Villa of Poggio Imperiale; to Prof. Bonelli-Righini and Prof. Fossi for their kind hospitality during the social events held in the Museums of the History of Science and of the Ancient Florentine House.

Last, but not least, we thank the local institutions, in particular the Health Assessorship of the Comune of Florence, for their financial and organization support.

Firenze, Italy, March 1980   *R. Pratesi   C.A. Sacchi*

P. Burlamacchi: "Laser Sources", in: *Lasers in Biology and Medicine*, ed. by F. Hillenkamp, R. Pratesi, C.A. Sacchi (Plenum, New York, London 1980)

X

Contents

---

* = Invited

Part I

General Introduction to Photomedicine and Photobiology

Photomedicine: Potentials for Lasers. An Overview

J.A. Parrish

Department of Dermatology, Harvard Medical School, Massachusetts General Hospital
Boston, MA 02114, USA

Photomedicine is one of the oldest and newest interests of
man. Belief in the health-giving properties of sunlight grew out
of sun worship, augmented by an increasing awareness that all
life on earth ultimately owes its existence to the sun. Since
ancient time, heliotherapy has been prescribed for a wide variety
of illnesses. Before the introduction of specific chemotherapy,
it was the mainstay of treatment for tuberculosis and other
chronic illnesses. In the nineteenth century, a scientific
rationale for such treatment arose with the discovery that microorganisms in vitro could be killed by ultraviolet radiation. The
1903 Nobel Prize for Medicine was awarded to Finsen for his work
demonstrating beneficial aspects of ultraviolet radiation of
cutaneous tuberculosis. Sunlight and artificial ultraviolet
sources were found to be useful in prevention and treatment of
rickets. With most of these human ailments, specific chemotherapy has replaced the use of phototherapy. However, UV lamps
are installed routinely in Siberian schools to prevent rickets,
and UV radiation is used to maintain sterility in operating rooms.

But now in the past two decades, a new gender of photomedicine
is emerging. The new science of photomedicine is based on
molecular photobiology, practical observations and scientific
method, and creative use of physics and sophisticated electro-
optical capabilities. Ultraviolet radiation and visible light
can now be used to prevent brain damage in newborn infants,
successfully treat chronic common skin diseases, prevent certain
forms of blindness, perform bloodless surgery and diagnose and
treat a variety of human diseases. We are learning about the
optical properties of skin and how to modify these properties.
We are learning details of photobiochemistry and how to selectively manipulate reactions in vivo. We are describing and
quantifying the in vivo effects of ultraviolet on blood cells,
the immune system and cancer cells, using this information to
devise schemes to improve health and to prevent and treat disease.
Technology has provided many new tools for photomedicine. Often
by the time the biologist or physician is able to formulate his
needs in a precise manner, the physicist and engineer have already
made available the equipment to meet those needs. The laser is a
unique new source of electromagnetic radiation and will continue
to markedly increase our capabilities in photomedicine. Its
present applications in surgery, dermatology and ophthalmology
serve as examples. Lasers emit photons which may react with

biological molecules to produce photochemical reactions and subsequent biologic sequelae. Photochemical and photobiological events are often the same, whether the photons come from the sun, a conventional light source or a laser. However, the molecular milieu, quantum yield, kinetics and photoproducts can also be quantitatively and qualitatively altered by cumulative properties of the laser: monochromaticity, coherence, and extremely high power density for very brief times. The limits of the application of lasers to photomedicine is not yet known.

A time of rapid growth of knowledge, skills and data is a time when overview and communication are most needed but also most difficult to provide. Laser scientists, engineers and physicists need to understand the questions of the biologist and the biologist must in turn comprehend the unique properties of available lasers. We begin here with reviewing some principles and observations regarding interaction of nonionizing electromagnetic radiation and human tissue. Then we will survey some of the active areas of photomedicine concentrating on therapeutic applications using both conventional and laser sources. Finally, we will hypothesize about the future of photomedicine and suggest areas in which the laser may find additional uses in research and treatment.

Only radiation which is absorbed results in photochemistry. As with any matter, x-rays, γ rays and other very high energy photons affect the highly organized and complex human tissues by relatively indiscriminate ionization of molecules, depending on density of electrons. The ionized molecules are highly reactive, and bonds may be broken or formed, but since the absorption is relatively nonspecific, the ability to select "target" molecules is limited. Infrared photons and microwaves excite specific vibrational or rotational modes, and therefore affect certain target molecules. However, the most significant biological effect of these wavebands is the heating caused by such kinetic excitation. The energy of photons in the ultraviolet and visible wavelengths is sufficient to cause electronic excitation of specific chromophore molecules, leading to specific chemical reactions. Thus, use of UV and visible radiation potentially offers the possibility of selecting among a wide variety of specific target molecules and photochemical reactions. Human color vision is an excellent example of how specific and biologically significant such processes may be.

The electronically excited states created by photon absorption may decay by two general processes, namely radiative and nonradiative loss of the excitation energy. Included in the latter is the formation or breaking of chemical bonds. Secondary reactions may occur. The results of these events on living cells may be inconsequential or they may have profound effects on metabolism, structure or function depending on the biologic importance of the chromophore molecules affected, and the dose of radiation. Nonionizing electromagnetic radiation penetrates into human skin, blood and eye and is absorbed by a variety of biomolecules. The absorption of radiation from the ultraviolet and visible range may result in _in vivo_ photochemistry which may lead to significant biological effects altering structure, function and lifespan of cells.

The skin is a large and important organ. Since it makes up about 15 percent of the total body weight, it is considered to be the largest organ of the body. In an adult, it is as if it were a living tissue system 6 feet long and 3 feet wide (1.8 square meters). The skin is alive. The outer layer of skin is continuously growing throughout life. Cells are constantly dividing and moving outward to replace the dead cells which fall off the surface. Any interference with this unending process leads to the problems we recognize as skin diseases.

Human beings are relatively naked. They are distinguished from most other land mammals by a lack of insulating fur. Instead, they have developed a unique combination of features: thick outer layers of skin with a well-developed dead horny layer, a widespread system of thermal-sensitive sweat glands, and an extensive layer of thermally insulating fatty tissue at the undersurface of the skin. This complex arrangement allows humans to survive in a wide range of temperatures and humidities. The purpose of the skin is to protect the host from a noxious environment and to maintain a homeostatic internal milieu. The skin absorbs much of the mechanical stresses of our world and also shields us from chemicals, sunlight, and bacteria. Acting as an insulator and a selective membrane, the skin keeps the environment within the body at a relatively constant temperature and saltwater content. The skin has a vast network of nerve endings which mediate the sensation of touch, heat and cold. When the skin performs these life-supporting functions normally, we notice it only for its aesthetic qualities. But if any of the protective mechanisms malfunction or become overwhelmed, we suffer embarrassment, discomfort, disfigurement, and possibly death.

Figure 1 is a diagrammatic cross-section of the skin showing 3 major tissue layers: the epidermis, the dermis, and the subcutaneous tissue. The thin, outermost epidermis is composed of tightly packed sheet of cells called keratinocytes beneath a very thin but tough outer layer called the stratum corneum. The stratum corneum provides protection against water loss and surface abrasion and attenuates ultraviolet radiation before it reaches living cells. The dermis is much thicker than the epidermis, has fewer cells, and is mostly connective tissue, which provides the substance, strength, and elasticity of the skin. Blood vessels, lymphatics, and nerves course through the dermis. The deepest layer, the subcutaneous tissue, is mainly fatty tissue and acts as an insulator and a shock absorber. The thickness of skin varies from one body region to another. Most of this variation is accounted for by differences in the thickness of the dermis. The epidermis is relatively uniform in thickness over the body except on the palms and soles, where it is much thicker.

Epidermis: The epidermal keratinocytes are so named because they produce keratin, the fibrous protective proteins of the skin. The exact structure of keratin is unknown. Keratinocytes are derived from a single germinative layer called basal cells (Fig. 2).

These stem cells divide causing some daughter cells to be pushed outward. The daughter cells which move toward the surface no longer divide but differentiate to form the precursors of keratin. As differentiation and outward migration continue, the keratinocytes lose their nuclei, dehydrate, and flatten out into dead, polygonal cells with a surface area about 25 times that of the basal cells (Fig. 3). This closely packed and cemented flat, dead cell layer laden with keratin and lipids forms the tough stratum corneum.

Specialized cells called melanocytes (see Fig. 2) reside between basal cells in the innermost cell epidermis. These cells produce melanin, a complex macromolecular protein derived from tyrosine which strongly absorbs visible light and ultraviolet radiation. Dendritic processes of the melanocytes interdigitate between keratinocytes and facilitate the transfer of melanin containing granules, called melanosomes, into the keratinocytes. These melanosomes are carried outward within the keratinocytes, and ultimately some melanin is deposited in the stratum corneum. Differences in skin color are mostly due to varying degrees of productivity of melanocytes, not the number, density or size of melanocytes. Absorption of UV radiation by melanin in the stratum corneum provides some protection against actinic damage to the skin; the tendency to sunburn or to have certain skin cancers is inversely related to how much melanin is present. Increased production of melanin (tanning) is induced following sufficient exposure to ultraviolet radiation.

Dermis: The dermis is mostly a semisolid mixture of fibers, water, and a viscous gel called ground substance. Ground substance consists largely of water and mucopolysaccharides. There are three types of fibers present: collagen, reticulum, and elastin. Collagen, which constitutes about 70 percent of the dry weight of dermis, is a long molecule woven into fibrils. Scattered cells called fibroblasts produce the fibers, proteins, and viscous materials of the dermis. The dermis is tough and strong. The complex nature of the dermis and its fibers creates a tissue with very high tensile strength which can resist compression but, at the same time, remains pliable and movable. Leather is animal dermis which has been modified by dehydration and certain processes (tanning) which render it stable and resistant to decomposition or bacterial decay.

Optical Properties of Skin and Blood: The complex structure of skin layers and the presence of appendages such as hair follicles, sweat glands, and sebaceous glands makes precise modeling of the path of optical radiation within the tissue difficult. A schematic representation of the optical pathways of radiation in skin is shown in Fig. 4. Within any given layer, four basic optical processes may occur:

1. Direct reflection at the boundaries of the layer due to change in index of refraction (greatest at the air/stratum corneum interface);

2. Scattering by molecules, particles, fibers, organelles, and cells within the layer;

Fig. 1.

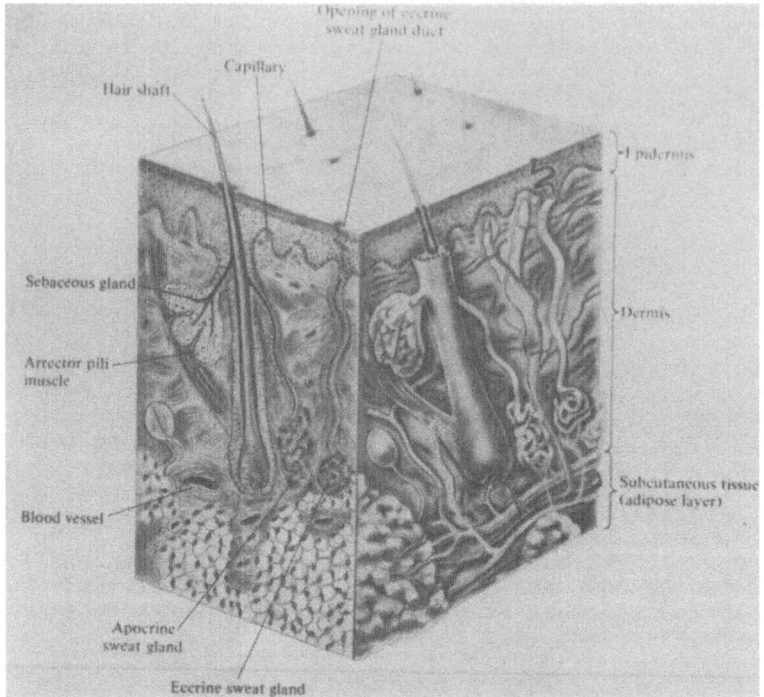

Opening of eccrine sweat gland duct
Capillary
Hair shaft
Epidermis
Sebaceous gland
Dermis
Arrector pili muscle
Subcutaneous tissue (adipose layer)
Blood vessel
Apocrine sweat gland
Eccrine sweat gland

Fig. 2.

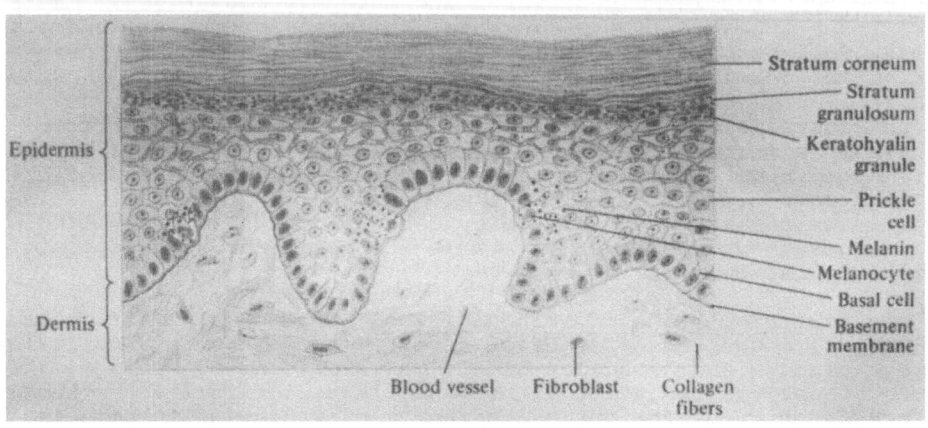

Epidermis
Dermis
Stratum corneum
Stratum granulosum
Keratohyalin granule
Prickle cell
Melanin
Melanocyte
Basal cell
Basement membrane
Blood vessel
Fibroblast
Collagen fibers

Fig. 3.

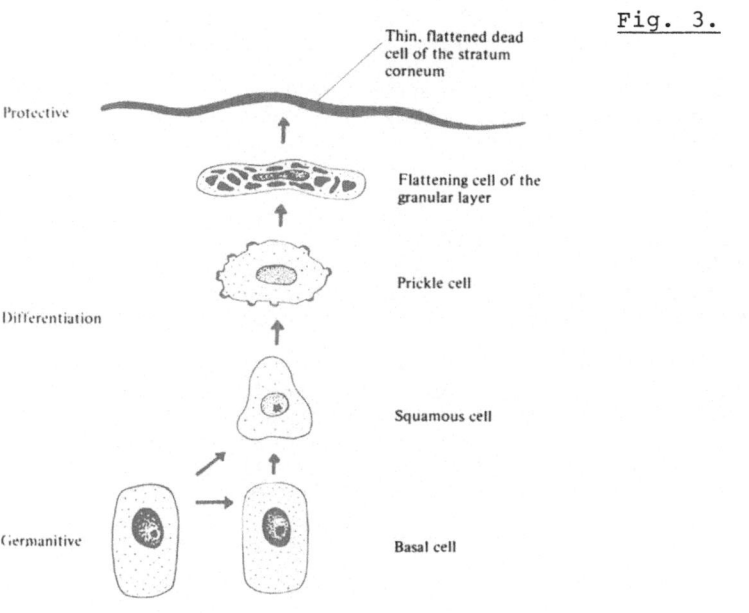

Thin, flattened dead
cell of the stratum
corneum

Protective

Flattening cell of the
granular layer

Prickle cell

Differentiation

Squamous cell

Germanitive

Basal cell

Fig. 4.

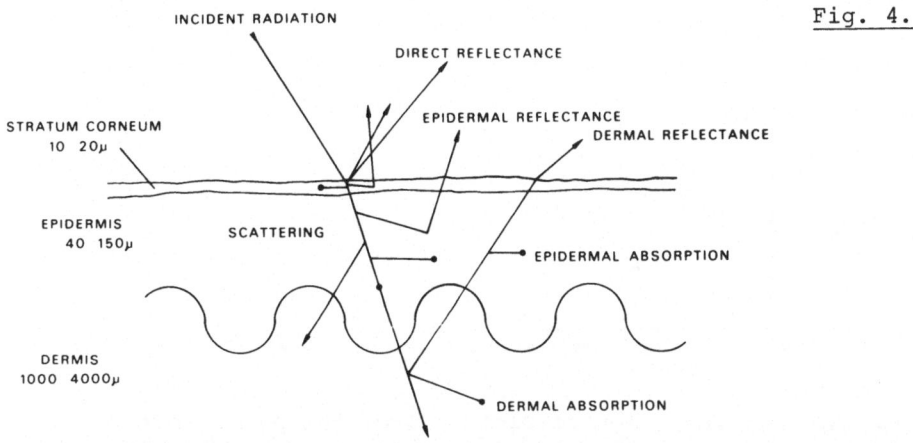

INCIDENT RADIATION

DIRECT REFLECTANCE

EPIDERMAL REFLECTANCE

DERMAL REFLECTANCE

STRATUM CORNEUM
10 20µ

EPIDERMIS
40 150µ

SCATTERING

EPIDERMAL ABSORPTION

DERMIS
1000 4000µ

DERMAL ABSORPTION

3.   Absorption (which may lead to photochemistry or dissipation
     of the absorbed energy via heat, fluorescence, or phosphor-
     escence); and

4.   Direct transmission through the layer.

     When radiation strikes the skin, part is remitted, part is ab-
sorbed in various layers, and part is transmitted inward to successive layers of cells, until the energy of the incident beam
has been dissipated. A very small fraction of the absorbed radiation is re-emitted at longer wavelengths (fluorescence). Between
5 and 7 percent of perpendicularly-incident light is reflected by
the outer surface of the stratum corneum because of the difference
in refractive index between air and skin. This surface reflec-

tance is relatively constant for all visible wavelengths and
therefore does not, in general, greatly affect our perception of
the color of skin. It does, however, account for the surface appearance of the skin. Fair Caucasian skin remits about half of
incident visible and near-infrared radiation. Most of this remission results from backwards scattering by collagen within the dermis. In passing through the overlying epidermis, however, visible
radiation is absorbed by melanin. Our perception of skin color is
determined by the visible remittance, "white" skin remitting more
than pigmented skin. Ultraviolet radiation of wavelengths less
than 320 nm is mostly absorbed within the stratum corneum or epidermis. In general, from 280 nm in the ultraviolet to 1000 nm in
the infrared, the longer wavelengths penetrate deeper into the skin.

There are many chromophores in skin and blood (Fig. 5). Proteins, amino acids, DNA, RNA, hemoglobin, bilirubin, and many
other small molecules and macromolecules absorb ultraviolet and
visible radiation. Some of these chromophores deserve special
attention from biologists because they give skin its color, protect the host from radiation or are involved in photochemistry
which leads to cell injury.

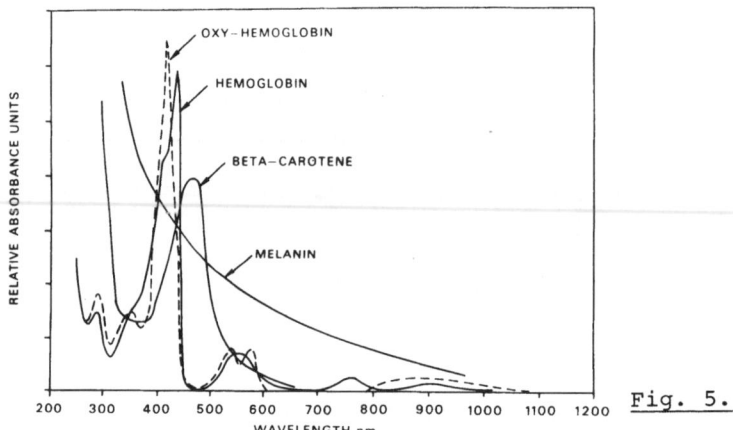

Fig. 5.

Ultraviolet Radiation of Blood: From the point of view of
metabolic need, the skin is vastly overperfused. The mean flow
is 20 to 30 times greater than the minimum necessary flow. This
difference is because cutaneous blood flow serves as heat regulator
of the entire organism and is not governed solely by metabolic requirements of the organ. Depending on body and ambient temperature, as much as 10 percent of blood is in the superficial vessels
of the skin available for ultraviolet exposure. Prolonged exposure
may make it possible to radiate a larger portion of blood and
blood cells as it courses through superficial skin vessels. Effects
of ultraviolet radiation on immunity and blood-borne metabolites
have been demonstrated in animals and humans.

Response of Normal Skin to Ultraviolet Radiation

Physicochemical molecular events occur immediately upon absorption
of ultraviolet photons within the tissue. These events may be

followed by alterations in biochemistry and subsequent changes in cell metabolism. Alterations of bases in DNA are the lesions which have received the most attention. This is because of the importance of DNA in cell regulation and replication, the relative absence of redundant molecules and the relative stability of some of the DNA photoproducts making them available for study. But it must be remembered that many other photochemical changes occur in RNA, proteins including enzymes, lipoproteins, and membranes. Some of these altered molecules may have little effect on the cell but others may alter metabolism or survival or lead to the release of chemicals that later affect adjacent cells or tissues.

During the hours after radiation, synthesis of DNA, RNA and proteins is decreased, metabolism is altered and histochemical evidence of cell injury is present. It is early in this period of decreased macromolecular synthesis that DNA repair occurs. Subsequently, the cell recovers, mutates or dies. Expression of mutation or death may be immediate or delayed by several cell cycles. Electron microscopic and histochemical studies demonstrate changes in cell structure and function occurring minutes to hours after exposure to ultraviolet radiation. Subsequently, products of photochemistry and altered metabolism lead to cellular changes recognizable by routine light microscopy over days to weeks. Over hours to days, changes in blood flow, cell kinetics, and pigment production cause grossly observable changes in the whole organ. Many of the immediate and most of the subsequent intracellular events remain unknown.

The effects of killing individual partially differentiated keratinocytes within the epidermis must be considered in view of the normal structure and function of the skin. The most superficial living epidermal cells are the most exposed and they are a population of cells that are already programmed to die as a final stage of their differentiation in becoming stratum corneum. Early death, abnormal differentiation or malfunction of a portion of these cells may not create significant problems to the host. Presumably, mutation of the stem cells (basal layer) that supply the epidermal cell population is of greater concern.

The response of the skin to ultraviolet exposure is, in general, a reparative and protective reaction. Sunburn is an example of inflammation, a generalized, primitive, protective, pathophysiologic response designed to remove injurious agents which invade an organ. In this sense, many features of the response are similar to those caused by other irritating or toxic agents. However, ultraviolet exposure is one cause of inflammation in which no matter actually enters the tissue, while producing little heating and no physical trauma. The injurious agents of ultraviolet exposure must be certain photochemical products formed within the tissue, which may include photochemically altered DNA, RNA, proteins, membranes and other molecules or structures of biologic importance.

Erythema (redness) results from dilation of blood vessels and is one aspect of the inflammatory response. The time course of erythema is variable and wavelength dependent. In normal skin, it is delayed in onset, usually peaks at 12-20 hr after exposure, depending upon the wavelength of the radiation, and may last several days. Because delayed erythema is the earliest, and some-

times the most striking, clinically observable ultraviolet-induced
alteration of skin, is in some way related to degree of damage,
and can be measured noninvasively, it is the most studied feature
of acute ultraviolet injury. It is the end point most commonly
used to determine the tolerance of normal skin to a single exposure of ultraviolet radiation.

The presence and degree of delayed erythema induced by exposure
to ultraviolet radiation is relative to the radiant energy delivered per unit area of skin surface and not to the rate of delivery
(irradiance) per se. For a given area, the exposure dose equals
the product of irradiance and exposure time. Within extremely
wide limits (10^{11} fold range) delayed erythema response is independent of irradiance ("dose rate"); that is, the influence of dou-
bling the irradiance may be compensated for by a halving of the
exposure time. This relationship, of a given exposure dose
yielding a constant biologic response, is termed the reciprocity
law. Lasers may be useful to study the limits of reciprocity in
vivo. Reciprocity over a wide range of irradiances is not the
case for thermally induced photobiologic responses in the skin.
As a site is exposed, heat transfer away from the exposure area
occurs until either some equilibrium temperature is reached at the
site during exposure or the exposure ceases. At equilibrium, the
temperature attained at the exposure site is proportional to the
irradiance, and not the exposure dose, and a deviation from the
reciprocity law is seen when the biologic response is affected in
some way by the temperature produced during exposure.

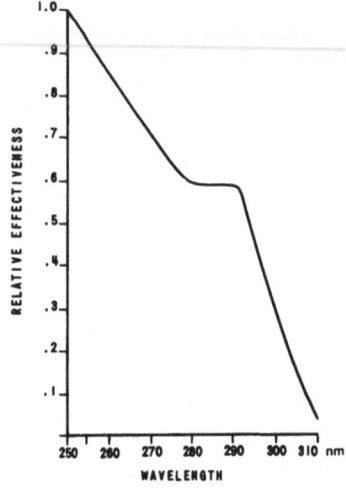

Fig. 6.

Action spectra in many cases are determined by the absorbance
of the chromophore molecules responsible for initiating the response chain. The complex optics and generality of the erythema
response, however, limit our ability to decide which initial
chromophores lead to erythema. One obtains the action spectrum
for delayed erythema by plotting the reciprocal of the exposure
dose necessary to produce a certain predetermined grade of redness
versus wavelength. The exact shape of the erythema action spec-

trum is still disputed and is affected by the degree of erythema that is being plotted, the time of reading after irradiation, and other variables. In general, the shorter wavelength ultraviolet photons are much more erythemogenic to normal human skin than are the longer wavelength photons (Fig. 6). At longer wavelengths, however, the dose-response curve rises more sharply, radiation is transmitted deeper within the skin and different chromophores and pathophysiology may be involved.

Melanin pigmentation is generally regarded as a major defense of the skin against the acute and chronic effects of sun exposure. Constitutive pigmentation describes the individual's baseline color and facultative pigmentation is the ability to tan in response to ultraviolet exposure. These characteristics are both genetically determined. Thickening of the epidermis is a generalized protective response often associated with inflammation. Increased proliferation of keratinocytes following ultraviolet exposure of skin leads to thickening of the epidermis and subsequently may be associated with a noticeable increase in desquamation or "peeling."

The only known beneficial effect of ultraviolet radiation on skin is the conversion of 7-dehydrocholesterol to previtamin D_3 in the skin, which then thermally isomerizes to vitamin D_3. Vitamin D_3 affects calcium absorption and calcification of bone, and insufficient solar UVB exposure causes rickets in children. Increased calcium absorption occurs in vitamin D_3-deficient subjects after ultraviolet exposures. While vitamin D_3 is produced within the skin by photochemical alteration of 7-dehydrocholesterol, hydroxylation of vitamin D_3 to 25-OH cholecalciferol (25-OHCC) in the liver and hydroxylation of 25-OHCC to 1,25 dihydroxychole-calciferal (1,25-OHCC) in the kidney are largely responsible for calcium absorption and homeostasis.

Long-Term Effects of Ultraviolet Radiation of Human Skin: Chronic effects of ultraviolet radiation include skin cancer and actinic alteration of skin. There is excellent epidemiologic evidence supporting the role of sunlight in human skin cancer. Basal cell carcinoma (BCC) is the most common skin cancer in Caucasians. These tumors are found primarily on the exposed surfaces of the head and neck. Patients with BCC generally have light eyes, fair hair, light complexions, sunburn easily, and spend more hours outdoors than do matched control people without such tumors. BCCs are rare in highly-pigmented races.

Squamous cell carcinoma (SCC) is the second most common skin cancer in Caucasians. The evidence for the role of sunlight in producing SCC is even more convincing than it is for BCC formation. SCCs are distributed primarily over the head, neck, and exposed areas of the upper extremities. They are found most abundantly in men who work outdoors. Both BCCs and SCCs are more prevalent in geographic areas of high sun exposure. The influence of latitude of residence on the mortality from melanoma in whites is also evidence for sunlight as a factor in induction of malignant melanoma.

The action spectrum for photocarcinogenesis between 250 and 420 nm has been tested using various broadband sources and appears somewhat similar to that for induction of delayed erythema. However, unlike erythema production, UVB is more efficient than UVC in producing tumors. In addition to being a known mutagen, UV radiation has been shown to affect the immunologic response to tumor cells. The mechanisms for UV-induced cancers are therefore thought to be somewhat complex.

The appearance of the chronically sun-exposed skin of Caucasian individuals is often referred to as either actinic (sunlight-induced) damage or "premature aging." This term is used because most of the features -- e.g., wrinkling and yellowing -- are similar to those observed in the process of "natural aging." It should be noted that chronic sun-induced damage produces gross and histologic changes, such as solar keratosis and solar elastosis, that are not necessarily the same as those seen in the process of natural aging. However, the difference may be one of degree. Susceptibility to actinic damage is genetically determined, light-skinned Caucasians being most prone while blacks are most resistant. While skin color and ability to produce melanin in response to sunlight exposure are the main factors involved, other genetic factors, such as the ability to repair the light-induced damage, may also be operative.

Terminology [1]: The ultraviolet radiation region of the spectrum has been subdivided into several bands in terms of phenomenologic effects. The subdivisions are arbitrary and differ somewhat depending on the scientific discipline involved. Dermatologic photobiologists generally divide the ultraviolet spectrum into 3 portions, called UVA, UVB, and UVC, in order of decreasing wavelength. In this text, the wavelength range from 200 to 290 nm is called UVC. Radiation of wavelengths shorter than 200 nm is mostly absorbed by air, and solar radiation of wavelengths below 290 nm does not reach the earth's surface because of absorption by ozone formed in the stratosphere. The band from 290 to 320 nm is called UVB, and the band from 320 to 400 nm is called UVA. The division between UVC and UVB is sometimes chosen as 280 nm, and 315 nm is sometimes chosen as the division between UVB and UVA. Because the divisions between UVC, UVB, and UVA are neither phenomenologically exact nor agreed upon, for critical work one should always define ultraviolet radiation in more rigorous spectroradiometric terms.

Radiation in the UVC band causes erythema of normal skin very efficiently and can cause photokeratitis (inflammation of the cornea). UVC is also called germicidal radiation because of its effectiveness in killing one-celled organisms. UVC is often called shortwave UV because the wavelengths in this region are the shortest ultraviolet radiation transmitted through air. This ultraviolet region is the furthest from the visible spectrum and is also called far UV.

Solar ultraviolet radiation of wavelengths between 290 and 320 nm reaches earth in relatively small quantities but is very efficient in causing sunburning of human skin. For this reason, it is often referred to as the sunburn spectrum or the erythemal band. Because of its relative spectral position, UVB is also called middle UV, mid UV, or middlewave UV. Radiation of wavelengths longer than 320 nm is relatively inefficient at causing redness of human skin. The UVB portion of the spectrum has been shown to in-

duce skin cancer in laboratory animals and mutations in bacteria.
Epidemiologic evidence suggests strongly that solar UVB causes
skin cancer in man. Long-term UVB exposure is also thought to be
at least partly responsible for producing the changes of exposed
human skin that are commonly termed aging.

Much less attention has been paid to ultraviolet radiation of
wavelengths longer than 320 nm (UVA, 320-400 nm). UVA is both
melanogenic and erythemogenic but the amount of energy required to
produce an effect is orders of magnitude higher than for the UVB
region. UVA is sometimes referred to as longwave UV and long UV
and is also called near UV because of its proximity to the visible
spectrum. This spectral region has also been called the blacklight
region, because its principal use for many years was to excite
fluorescent and phosphorescent substances that reradiate the absorbed energy as light in the visible spectrum. UVA has recently
received more attention [1] because:

1. High-intensity sources of UVA including lasers are now
   available,
2. UVA has been shown to affect cells and microorganisms,
3. UVA may potentiate or add to the biologic effects of UVB,
4. Photosensitivity reactions (phototoxicity and photoallergy)
   are mostly mediated by UVA,
5. There is experimental and epidemiologic evidence to suggest
   that solar UVA is one of the possible etiologic agents for
   certain kinds of cataracts in humans,
6. UVA-induced photopolymerization and photochemical reactions
   are being used in industry to alter rubber, plastic, glass,
   metal, paper and photographs. Photopolymerization of certain chemicals provides convenient ways to apply dental and
   orthopedic appliances and "photocure" them in place, and
7. The use of UVA in conjunction with photosensitizing drugs
   has opened up new therapeutic possibilities in chronic skin
   disorders, such as psoriasis, mycosis fungoides, and eczema.

Abnormal Responses to Ultraviolet Radiation [2,3]

Abnormal responses to ultraviolet and visible radiation are many
and include a special challenge to the photobiologist. Chemical
photosensitivity is a term which embraces all forms of photosensitivity resulting from excitation of an identified chemical by
electromagnetic radiation. A wide variety of photosensitizing
chemicals of therapeutic, industrial, agricultural, or other origin
may reach the skin directly or via the bloodstream, each having
its own pattern of absorption, metabolism, and binding to skin
components. A variety of photochemical and molecular mechanisms
are involved. In general, the compounds possess highly resonant
structures with a molecular weight of less than 500 and absorb
radiation in the ultraviolet and visible range. Chemical photosensitivity has been traditionally divided into two clinicopatho-
logic classes, phototoxicity and photoallergy. Photoallergy is
photosensitivity mediated by immunologic pathways. Light-induced
damage that is not dependent on an allergic mechanism may be considered phototoxic in nature.

The porphyrias are disorders of porphyrin or porphyrin precursor metabolism some of which result in the production of excess amounts of photosensitizing chemicals which when present in skin and blood lead to abnormal sensitivity to light. Polymorphous light eruption is a common photodermatitis of unknown etiology which exhibits a variety of morphological appearances (erythema, eczema, papules, plaques, blisters) on light-exposed areas. Solar urticaria consists of wheals appearing within seconds or minutes of exposure to sunlight. Solar urticaria is evidently a syndrome that may result from several different pathogenic mechanisms.

Nonvisual Photobiology of the Eye

The eye is a complex organ designed to convert electromagnetic energy into chemical energy and into neuronal impulses which, when processed in the brain, provide an instantaneous map of our environment. Ultraviolet radiation is absorbed by the anterior components of the eye and can lead to inflammatory responses. Wavelengths shorter than approximately 290 nm are partially or completely absorbed within the cornea and conjunctiva. The acute effects of excessive exposure to these wavelengths are conjunctivitis and a corneal inflammatory reaction known as photokeratitis. After ultraviolet exposure, there is a period of latency varying somewhat inversely with the amount of exposure. The latent period may be as short as 30 min or as long as 24 hr, but it is typically 6 to 12 hr. Conjunctivitis is associated with the sensation of a foreign body or "sand" in the eyes, varying degrees of photophobia, lacrimation, and spasm of lid muscles. Corneal pain can be very severe. Acute symptoms usually last from 6 to 24 hr, and almost all discomfort disappears within 48 hr. Very rarely does exposure result in permanent damage. Unlike the skin, the ocular system does not develop tolerance to repeated ultraviolet exposure.

Ultraviolet wavelengths longer than 290 nm may reach the lens, the iris, and the aqueous humor (anterior chamber). Experimental evidence in rabbits and monkeys indicates that permanent lenticular cataracts may be produced by single, high-irradiance or long-exposure durations to UVB or UVA. This damage may be induced by thermal and/or photochemical mechanisms in the lens. In albino mice, cataracts may be produced by multiple daily UVA exposures that are below the single-exposure threshold dose of observable corneal damage. Ultraviolet radiation may alter lens crystalline proteins from soluble, lower-molecular-weight crystallines to insoluble, higher-molecular-weight crystallines which may cause light scattering within the lens. Tryptophan photochemistry appears to be involved in some forms of ultraviolet induced cataracts.

Focal photocoagulation phototherapy of the posterior ocular tissues using xenon arc sources has been practiced since 1950. Lasers are now rapidly becoming the principal tool used for selective absorption by hemoglobin or melanin to induce focal photocoagulation or scarring to seal or wall off retinal detachment or retinal tears, reduce macular edema by occluding leaking retinal vessels, and destroying microaneurysms and neovascular tufts. Laser surgery has also been used in ophthalmology.

Photomedicine

The general scope of photomedicine [2,3] is the manipulation of
exposure dose parameters and the alteration of the host with exogenous agents to maximize the beneficial and/or minimize the adverse effects of nonionizing electromagnetic radiation. This
usually necessitates identification and quantification of the effects of that radiation on normal and abnormal human tissue and
the study of molecular and cellular mechanisms of photopharmacology
and pathophysiology. The study, diagnosis, treatment and prevention of photodermatoses, skin cancer and chronic actinic change of
skin are important tasks within photomedicine. One specific aspect
of photomedicine is the use of nonionizing electromagnetic radiation, with and without the addition of photoactive drugs, to treat
disease by altering cell metabolism and extracellular metabolites.
It is this aspect of photomedicine which we will now survey.

Phototherapy of Uremic Pruritus [4]: Generalized pruritus is
one of the most common and most bothersome symptoms of chronic
renal failure affecting up to 85% of patients on hemodialysis
maintenance. Conventional treatment of itching is rarely helpful.
It has been found that 6 to 8 exposures to erythemogenic doses of
UVB markedly reduces the severity of uremic pruritus. Placebo
exposure (suberythemogenic doses of UVA) has no such effect. UVB
phototherapy of one side of the body results in generalized improvement without localization of the beneficial effect to the
exposed side. This suggests that the therapeutic effect results
from the inactivation of a circulating substance present in uremia
and responsible for the itching experienced in persons with chronic
renal failure.

Phototherapy of Hyperbilirubinemia [5]: Bilirubin, a metabolite
of heme breakdown is normally made water soluble by hepatic enzymes and then is excreted into the gut to be eliminated. Glucuronyl
transferase, the hepatic enzyme which facilitates the "conjugation"
of bilirubin to its more water soluble form, is not maximally
active at birth. In newborns, especially in premature infants,
unconjugated bilirubin may accumulate in the blood and tissues.
The appearance of this excess metabolite in the skin leads to
jaundice and the accumulation in neonatal brain tissue may cause
irreversible brain damage. Exposure of the skin to blue light
decreases free bilirubin in the blood and skin and decreases the
incidence of central nervous system sequelae. While it is known
that bilirubin can be photodegraded by photooxidative processes in
vitro this is probably not the most important therapeutic mechanism
in vivo. Photoisomerization of unconjugated bilirubin results in
a more water soluble form which can be passed directly into the
bile for elimination through the gastrointestinal tract [6]. The
fact that unconjugated bilirubin appears in the bile in increased
amount after phototherapy of jaundiced humans and hyperbilirubinemic experimental animals supports this mechanism of beneficial
effect. Phototherapy of jaundiced infants is widely used throughout the world.

Phototherapy of Skin Diseases: Repeated exposures to ultraviolet radiation results in improvement of certain dermatoses. The
exposure dose required is near that dose which, after a single
exposure, causes inflammation of uninvolved skin as manifest by

delayed erythema. It is assumed that after ultraviolet exposure
the same type of cascade of cellular events and inflammation that
occurs in normal skin occurs in diseased skin. It is further
assumed that repeated phototoxic cell injury results in a therapeutic effect by either altering abnormal cell kinetics, depleting
or altering mediators or metabolites, or selectively injuring
specific host cells which play etiologic or essential roles in
specific skin diseases. It is not known why skin disease often
remains in remission after phototherapy is discontinued. It is
possible that repeated induction of DNA lesions or decreased macromolecular synthesis is therapeutic in psoriasis, a hereditary
disease of epidermal hyperproliferation. In polymorphous light
eruption, phototherapy may induce protective melanogenesis, deplete
mediators or alter lymphocyte function. Lymphocytotoxicity may
also be a factor in phototherapy of certain forms of eczema or
mycosis fungoides. In disorders such as acne a major benefit is
the induction of tan which cosmetically masks the disease process.

Therapeutic Topical Photosensitizers: The two best known and
most widely used examples of the use of topically-applied photosensitizers to treat skin disease are "dye-light" therapy of recurrent herpes simplex and tar-ultraviolet treatment of psoriasis
(Goeckerman regimen).   The former has proven ineffective to date
and the latter may not work via photosensitization. Uncontrolled
clinical studies suggested that the symptoms of recurrent herpes
simplex (cold sores, fever blisters) were diminished by application
of neutral red or proflavin and subsequent exposure of the affected
skin to visible light. Photodynamic inactivation of the causative
virus was thought to be the therapeutic mechanism. When it was
discovered that surviving fractions of photoinactivated viral populations had oncogenic potential questions arose regarding the use
of this form of phototherapy. Subsequently, double-blind controlled studies failed to show definite benefit in humans.

The action spectrum for phototherapy of psoriasis roughly paral-
lels portions of the action spectrum for delayed erythema with or
without certain photosensitizers. Cell injury of normal skin is
the short-term limiting factor in phototherapy.

The relative therapeutic effects of tar-ultraviolet treatment
of psoriasis are complex and tar has some definite but minimally
therapeutic benefit of its own. Some of the many ingredients of
crude coal tar are photosensitizers with action spectra in UVA
but even in the presence of tars normal human skin is much more
sensitive to UVB radiation. Because of the spectral power distribution of the ultraviolet sources normally used in treatment, UVB
phototoxicity (manifest by delayed erythema) limits the therapy
and observable photosensitization by tar plus the UVA component
of the sources does not occur. While tar-UVA therapy is effective
it is impractical because of long treatment times and immediate
onset of a sensation of burning of skin. There is evidence that
the lubricant vehicle used to apply tar alters the optical properties of the psoriatic scale to selectively increase ultraviolet
transmission compared to the adjacent normal skin. This may increase the known therapeutic effect of repeated exposures to UVB.
The exact mechanism of phototherapy of psoriasis is not known.

Photochemotherapy: Photochemotherapy describes the combination
of drug plus nonionizing electromagnetic radiation to treat disease; the site and nature of the in vitro photochemistry and subsequent photobiology are altered by supplying an exogenous chromophore. In the doses usually used, the drug alone or the electromagnetic radiation alone have no effect. Hematoporphyrin derivative is a photosensitizer, appears to selectively remain in higher
concentration in malignant tissue, and has therefore been used to
treat metastatic cancer. The hematoporphyrin is also fluorescent,
and can be used to detect tumor masses. The requirements for
light sources for this form of photochemotherapy and diagnosis
make the laser a reasonable consideration. High intensity and
monochromaticity can be used to augment selective tissue destruction at varying depths in tissue. The coherence properties of
lasers, with its associated collimated beam, permits focusing of
high power levels into fiber optics, which may then be used to
both locate and treat internal neoplasms.

Recently the known potent photosensitizing properties of psoralens have been better quantified and utilized to treat a variety
of diseases, including psoriasis, mycosis fungoides, vitiligo and
some forms of eczema. The treatment, which is often called PUVA,
utilizes orally-administered psoralen (P) and subsequent exposure
of the skin to longwave ultraviolet radiation (UVA). The long-
term effects of this treatment are not yet known.

Photoimmunology [7]: Immune responses are remarkably versatile
adaptive processes in which the host forms specifically reactive
cells and proteins in response to an immense variety of biologic
macromolecules and organic molecules. This faculty was apparently
acquired late in evolution because classic immune responses are
found only in vertebrates for whose survival they are essential.
These acquired responses to foreign molecules are not only the
principal defense against infection by pathogenic microorganisms
and viruses but also constitute a major defense against host cells
that undergo certain mutations or malignant transformation. In
humans, immunology and photobiology appear to interdigitate in the
pathophysiology of certain photodermatoses (polymorphous light
eruption, solar urticaria), in the mechanism of certain forms of
phototherapy and photochemotherapy (eczema, mycosis fungoides) and
in the induction of skin cancer (UV-induced tolerance to UV-trans-
formed cells).

It has been known for over 50 years that exposure to ultraviolet radiation could abolish the function of certain antibodies.
This effect and other electromagnetic radiation induced alteration
of antibody function simply reflect in vitro photochemical and
chemical changes in proteins. Recently it has been observed that
ultraviolet radiation can affect cellular immunity in vivo. Macromolecules in the host may be altered photochemically in situ to
become partial or complete antigens. Subsequent formation of
antibodies by the host may be a helpful defense but may also have
pathophysiologic consequences. Under certain circumstances DNA
can be made to be measurably antigenic by in vivo ultraviolet
exposure. Possibly of more importance is the fact that the initial processors and final executors of cellular immunity reside
in the skin and blood and are therefore susceptible to ultraviolet and visible radiation. In both humans and animals the viabi-

17

lity and function of lymphocytes is altered by erythemogenic exposures of the skin to ultraviolet with and without the previous
administration of photosensitizers. Delayed hypersensitivity
reactions in skin can be suppressed by ultraviolet exposure prior
to or after initiating immune responsiveness. Photoimmunology
offers potential for studying mechanisms of disease and phototherapy and may prove to be a useful tool for manipulating immune
responses in humans.

Lasers in Photomedicine

At the molecular level, when individual photons of a given quantum
energy encounter biologic matter, usually the source of the quanta
(sunlight, arc lamp, blackbody, etc.) is unimportant. At the tissue level, because intensity-time reciprocity often holds over a
wide range and action spectra are generally broad, many different
conventional sources can be tailored to cause photobiologic responses which are very similar or have minimal quantitative differences. In animal and human skin it has been shown that an
ultraviolet laser can be used to give predictable, quantifiable,
reversible end points such as erythema and pigmentation and that
the energy and wavelengths required are generally the same as
those for conventional sources. However, the laser's high intensity, coherence and monochromaticity may make it a unique source of
photons so that under selected conditions molecular events and subsequent photobiologic responses are quantitatively and qualitatively different from those caused by conventional sources (Fig. 7).

LASER PROPERTIES	PRACTICAL ADVANTAGES	"QUALITATIVELY" DIFFERENT RESPONSES
High Photon Density ± very short pulses	Large Dose Effects	Two Photon Processes ↑ Thermal Effect Time Resolved Effects
Coherence collimated beam	Optical Path Focusing	
Monochromatic		Selective Photobiological Effects

High Intensity: The laser can produce uniquely high photon
densities and is capable of very short pulses of high peak power.
Lasers make it possible and practical to induce photobiologic
effects which require large energy doses. Some photobiologic responses require total exposure doses which can only be achieved
after minutes to hours of irradiation with conventional sources.
Such long exposure times are impractical for research or treatment,
and reciprocity may fail because repair and recovery processes may
be active during exposure. For example, delayed erythema induced
by longwave ultraviolet radiation requires 1000 or more times as

much energy than is required to induce erythema with more erythemogenically efficient shortwave ultraviolet photons. More importantly, the nature of the molecular and tissue events may be influenced by high photon density. Two photon processes may occur. The production of macroscopic thermal effects versus photochemically induced effects can be adjusted by varying peak and average power separately. In tissue, high average power may favor macroscopic thermal effects over photochemical-photobiological effects expected from conventional sources. The high peak powers attainable with lasers can produce microthermal effects at specific sites of absorption while sparing gross thermal damage. When these mechanisms are induced, intensity-time reciprocity no longer holds; photobiologic end points are not directly related only to total exposure dose. Considering that ultraviolet radiation may induce a complex cascade of chemical events and competing repair mechanisms with various limiting rates, appropriate changes in peak power and duty cycle may cause striking quantitative and qualitative differences in the tissue response. Thermal effects may occur with higher yield than photochemical effects expected from conventional sources. For these and other reasons, intensity-time reciprocity no longer holds; photobiologic end points are not directly related only to total exposure dose.

Coherence: The coherence property of lasers and the associated collimated beam facilitates directing the radiation through optical paths which would be difficult to accomplish with most conventional sources. The collimated beam can more conveniently be directed through fiber optic bundles or endoscopes which permit flexibility of exposure delivery and provides access to many sites within the human body. The collimated beam also permits focusing laser radiation with less dispersion than that found with conventional sources. This makes it possible to expose very small sites within tissue. It is possible to selectively radiate organelles or parts of living cells.

Monochromaticity: The monochromaticity of lasers, besides providing an ideal tool for the spectroscopist, also makes it possible to select chromophores from within complex biologic media or to isolate segments within wide photobiologic action spectra. Within complex biologic tissue, when absorption spectra of chromophores overlap, a monochromatic source can more easily select a wavelength unique to one of the chromophores. Subsequent photobiologic end points can then be selectively increased or decreased. This is especially important when considering lasers as therapeutic sources because wavelength selection may maximize benefit and minimize undesirable side effects. Also, by simultaneously radiating with two different monochromatic sources, waveband interactions could be utilized to obtain selective photobiologic responses.

Presently, lasers are especially useful in photomedicine as research tools. Spectroscopy, especially Raman scattering spectroscopy, are aided by the laser's monochromaticity and high power. Laser flow microphotometry is useful as a diagnostic tool. Cells are made to flow in a single file through a continuous wave laser beam and various optical characteristics are measured. The properties which can be measured include: presence of trypan blue dye or fluorescent stains, nuclear density and nuclear cytoplasmic ratio, presence of parasites or viruses in cells, differential

histochemical staining abilities. Quantification of these and
other properties makes it possible to measure cell cycle phases,
immunologic markers, malignant characteristics, viability and other
properties. Lasers have also aided the development of diagnostic
holograms, a procedure which can overcome depth of focus limitations of conventional photography. With this type of equipment
the internal structures of the eye can be studied and it may be
possible, using acoustical holography, employing the laser as a
means of detecting the diffraction patterns of ultrasound, to
obtain three-dimensional pictures of internal organs of the human
body. Lasers have also been employed to detect pre-clinical increases in scattering within the human ocular lens, which are a
sign of cataractogenesis.

Presently, the major medical uses of lasers involve controlled
thermal destruction of tissue. A focal temperature rise of several
hundred degrees can be achieved, resulting in vaporization and
explosive expansion. Tumors can be destroyed. Selective absorption can be used to destroy pigmented or blood-filled lesions.
Coagulation of vessels can be used to seal leaking or damaged
vessels. Scarring can be induced to confine or seal retinal de-
tachments or tears. By using appropriate photon density, duration
of exposure and wavelength, it is possible to use laser radiation
to destroy a thin line of tissue and coagulate the adjacent blood
vessels. The resulting bloodless incision is useful for many
forms of surgery.

The Future

Diagnostic and surgical uses of lasers will expand as a function
of growth in technology and increased knowledge of the optical
properties of skin and blood. In addition, lasers may be used to
induce nonlethal cellular changes and reversible tissue alterations
and may therefore be the appropriate exposure source for certain
forms of phototherapy and photochemotherapy. The major present
disadvantage of lasers for this use is their cost and limitations
in treating the whole surface area of human skin. Most forms of
phototherapy involve exposure of large areas of skin. If the
laser beam is spread or swept over large areas the average power
is often decreased well below that available with less expensive
sources. Coherence and impressive monochromaticity are not essential to most forms of phototherapy and photochemotherapy at present. However, our need for these characteristics may increase
substantially as we learn more about the mechanisms and side effects of phototherapy. There are several trends and concepts in
research which permit hypothesizing about future aspects of photomedicine which may involve laser radiation.

The concept of photochemotherapy will probably prove to be more
important than any specific example presently in use. Systemic
delivery of photoactive chemicals is convenient and achieves a
more uniform, predictable distribution than usually permitted by
percutaneous administration. In addition, the site of action of
the systemically-administered chemicals can be controlled at the
time of subsequent photoactivation by nonionizing optical radiation,
so that areas of the body not affected by disease can be shielded
and spared. The depth of therapeutic action is influenced by the
properties of the activating photons. Careful selection of the

chemical, wavelength, timing and intensity of exposure can produce specific effects on the viability and function of target cells, organelles, or molecules in selected body sites, including skin, blood and organs of the body that are accessible to optical radiation. Photochemotherapy is a treatment modality in which specific chemical reactions can be produced in a controlled manner at specified tissue depths, in selected body sites, using photons as the activation energy.

Possible therapeutic mechanisms include not only direct photodynamic or photosensitization participation of exogenous chromophores in producing altered biomolecules but indirect effects such as local photochemical activation or inactivation of enzymes, substrates or drugs, which may have profound non-photobiologic effects. Drugs or enzymes can be chemically bound _in vitro_ to side chains which render them inactive. If the linking chemical bond is designed to be subsequently broken by photons received _in vivo_ after the drug has been given to the host, the drug can be locally released at will.

Liposomes (microscopic vesicles made of lipid bilayer membranes) can be synthesized in a way that they trap chemicals within them. Subsequent heating of these particles to the specific liquid-crystalline transition temperature induces molecular rearrangements in the membrane that allows the trapped drug to leak out [8].

The transition temperature region varies with the lipid constitution of the liposome and can be a sharp transition within the physiologically tolerable temperature range. Liposomes filled with drug could be administered to the host and subsequent _in vivo_ exposure to conducted heat, infrared, or high irradiance visible or longwave ultraviolet radiation could cause local release of drug. The timing, site, and tissue depth of the drug release is influenced by the source of the local heating and distribution of liposomes. Such therapy could be termed thermochemotherapy. Liposomes could also be made to incorporate specific chromophores in their membrane. The _in vitro_ incorporation of drug plus subsequent _in vivo_ exposure to appropriate waveband of nonionizing electromagnetic radiation could rupture the liposome and lead to local drug release (photochemochemotherapy). If the synthesized liposomes contained a photoactive drug a second waveband could, subsequent to local release, activate the drug at specific sites. Therefore, thermophotochemotherapy and photochemophotochemotherapy become possible selective forms of therapy.

An area of promise is the study of waveband interactions and how these may influence phototherapy of all kinds. It is unwise to assume (as do most safety threshold envelopes) that radiation effects caused by various wavebands making up action spectra of human cells are always additive. In complex human organs action spectra may be made up of many competing, augmentive, synergistic and additive processes. Repair phenomena can be increased or decreased by previous, simultaneous or subsequent electromagnetic radiation within or outside of the primary action spectrum for a given photobiologic response. Infrared radiation may modify several aspects of the sequence of events following _in situ_ photochemistry. It is likely that photobiologic responses can be modified and risk/benefit ratio be improved by increased knowledge of

21

waveband interactions. By careful use of the various photon densities and pulse widths made available with lasers, it may be possible to induce responses which deviate from the usual laws of reciprocity but which do not kill cells or destroy tissue. If therapeutic responses and undesirable side effects differ in their dependence on intensity, manipulation of exposure dose parameters make it possible to make treatments more acceptable.

The potential for phototherapy, photochemotherapy and thermochemotherapy is greatly enhanced by growth in three areas of science: (1) advances in electronics, physics and optics have expanded the capabilities of optical diagnostics, photon dose delivery and quantitative measurements of photobiologic end points; (2) many photobiological principles have been thoroughly explored at the cellular and molecular level by scientists who work with plants, bacteria, photosynthesis, photochemistry and simple organisms with particular emphasis on reciprocity and wavelength interactions; and (3) increased understanding of the effect of nonionizing optical electromagnetic radiation on mammalian cells in vitro. The laser has and will continue to participate in all three of these areas. Moreover, lasers may potentially increase the effectiveness of present therapies and aid in the development of or be essential to new forms of treatment.

References

1. Parrish JA, Anderson RR, Urbach F, Pitts D: UV-A: Biological Effects of Ultraviolet Radiation with Emphasis on Human Responses to Longwave Ultraviolet. Plenum Press, New York, 1978, 262 p

2. Epstein JH: Photomedicine, in The Science of Photobiology (KC Smith, ed), Plenum Press, New York, 1977, pp 175-207

3. Parrish JA, White HAD, Pathak MA: Photomedicine, in Dermatology in General Medicine (TB Fitzpatrick, et al., eds), McGraw-Hill Book Co., Inc., New York, 1979, pp 942-994

4. Gilchrest BA, Rowe JW, Brown RS et al: Ultraviolet phototherapy of uremic pruritus. Ann Int Med 91:17-21, 1979

5. Odell GB, Schaffer R, Simopoulos AP: Phototherapy in the Newborn: An Overview. National Academy of Sciences, 1974, 190 p

6. McDonagh AF, Lightner DA, Palma LA: Photoisomerization of bilirubin in vivo. Presented at 7th Annual Meeting of the American Society for Photobiology, June 24-28, 1979, Pacific Grove, Calif., p 177 (Book of Abstracts)

7. Morison WL, Parrish JA, Epstein JH: Photoimmunology. Arch Dermatol 115:350-356, 1979

8. Yatvin MB, Weinstein JN, Dennis WH, Blumenthal R: Design of liposomes for enhanced local release of drugs by hyperthermia. Science 202:1290-1292, 1978

Common Misconceptions about Light

K.C. Smith

Department of Radiology, Stanford University School of Medicine
Stanford, CA 94305, USA

All biological responses to light are the result of chemical and/or physical changes induced in biological systems when they absorb light energy. The absorption of light energy by biological molecules, and the subsequent dissipation of this energy (via heat, light emission or chemistry) are well understood, and obey the laws of physics and chemistry (1). Unfortunately, not all of those who use light experimentally in the laboratory or therapeutically in the clinic are trained in the physics and chemistry of light. Thus, even among scientists and physicians there exist numerous misconceptions about the properties of light (1). The following are some of the more common misconceptions that I have observed.

1. Wrong: Visible light is natural and therefore is safe. There are two misconceptions in this statement. First, because something is natural is no reason to consider it safe. After all, poisonous snakes and plants are natural, but they certainly aren't "safe".

Second, the safety of light is not an intrinsic property of the light. The first law of photochemistry states that light must be absorbed before photochemistry can occur. Therefore, if the light is not absorbed by a system it is "safe" for that system. Visible light can be safe for one system, and extremely harmful to another. For example, blue light is "safe" for pure deoxyribonucleic acid (DNA) since it isn't absorbed by DNA, but is not "safe" for bilirubin since it is absorbed by bilirubin.

2. Wrong: Visible light is not as photochemically active as ultraviolet (UV) radiation. This misconception has arisen from two other misconceptions. It is true that biological systems are generally more easily inactivated by UV radiation than by visible light, but this is not due to the properties of the light, but rather to the properties of the light-absorbing molecules whose photochemical alteration leads to the inactivation of the biological system. Thus, since DNA, the most important molecule in a cell, can be altered by the absorption of UV radiation, a cell is most easily killed with UV radiation. This fact has contributed to the misconception that visible light is less photochemically active than UV radiation.

The photons of UV radiation carry more energy than do the photons of visible light. Since UV radiation is more effective in killing cells than is visible light, some people have incorrectly assumed that it is the absolute energy of a photon that is important in determining its photochemical potential. As stated above, unless a photon is absorbed by a molecule it cannot cause photochemistry. Thus, the energy of a photon is important only in that it be of the correct energy to permit its absorption by the molecular species of interest. The fact that photons of visible light carry less energy than do photons of UV radiation has also contributed to the misconception that visible light is less photochemically active than UV radiation.

It should be remembered that all wavelengths of light (UV through visible) are very active in initiating photochemical reactions, providing they are absorbed in appropriate molecules. No one wavelength is more photochemically reactive than another except as it relates to a specific target molecule. The absorption spectrum of a molecular species indicates what wavelengths of light are absorbed by that species.

3. <u>Wrong: We need not be concerned with the biological effects of visible light because it does not penetrate human tissues.</u> This misconception arises because of the further misconception that because we cannot see through a human hand, that light must not be transmitted through human tissues. When confronted with this incorrect statement, I ask if that person has played the childhood trick of putting the lighted end of a flashlight into his mouth or against his hand while in a dark room. Such a trick dramatically exemplifies that light does penetrate living tissue. It is mainly red light (670-760 nm) that is transmitted.

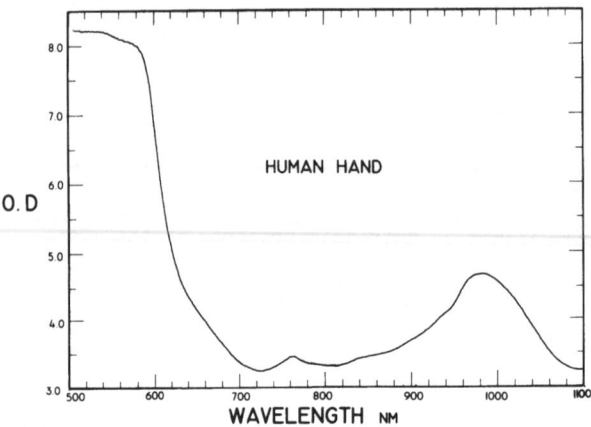

Fig.1 The absorption spectrum of a human hand. The spectrum was recorded by K.H. Norris with a very sensitive spectrophotometer with his hand close to the detector. The hand is rather transparent to light of wavelengths between 650 and 900 nm. An optical density (O.D.) of 3.5 corresponds to 0.05% transmission. (From reference 1.)

A dramatic example of the transmission of light by a human hand is on the cover of the March 1972 issue of <u>Scientific American</u>. The space between the fingers and the edges of the hand were carefully masked, and a flash lamp was fired on one side of the hand; a color photograph taken on the other side by the <u>transmitted</u> light yielded a picture of a red hand.

Using a very sensitive spectrophotometer, the absorption spectra of objects that seem opaque to human vision can be readily determined. The absorption spectrum of a human hand is shown in Fig. 1. A hand is rather transparent to wavelengths of light between 650 and 900 nm, however human visual acuity only extends between 380 to 700 nm, with a peak of efficiency around

550 nm. Thus, because humans can't see through the human body does not mean that the human body is opaque to all wavelengths of light.

Since light can penetrate deeply into human tissues (especially red light), and since absorbed light can cause photochemistry, it is appropriate to be concerned with the biological consequence of such absorbed light in the tissues of man.

4. Wrong: Light of different wavelengths always acts independently on biological systems. Actually when a biological system is exposed to two different wavelengths of radiation, the observed effect is frequently not the summation of the effects of the individual wavelengths (1,2). Most frequently one observes a synergistic effect, i.e., the effect of two wavelengths of radiation given together is greater than the sum of the two effects when given independently. Sometimes the effects are antagonistic (i.e., the opposite of synergistic), and often there is no interaction at all. Therefore, it is especially important to keep in mind the possibility of nonadditive effects of different wavelengths of light when one uses a polychromatic source of radiation, or two or more monochromatic sources.

5. Wrong: Laser radiation has magical properties. Lasers have long been the source of a great many misconceptions. Many people still feel that light from lasers has magical properties. Lasers can seem magical if their unique properties of micro-dot focusing, very high intensity, coherent radiation, possibility of ultrashort pulses, and monochromicity are made use of. If the first four properties are not useful in a particular application, then lasers are just expensive monochromatic light sources, whose emitted radiation follows (except for coherence) all the same laws of physics and chemistry that the same wavelengths of light from a conventional light source follow. Thus, in searching for new uses for lasers in the field of photobiology and photomedicine, one must make sure that the proposed application requires one or more of the unique properties of a laser. If this is not the case, then a conventional light source may be more cost effective.

One practice that has furthered the misconception of the magical properties of lasers is the use in publications of vague terms such as "laser radiation" or "sublethal ruby laser beams" instead of specifying the wavelength causing the effect. In many cases it may be irrelevant that a laser was used, but in all chemical and biological applications it is important to know the wavelength, the fluence rate, and the total fluence delivered to the target.

Also, before ascribing magical properties to laser radiation, the same studies should be performed with a conventional light source emitting the same wavelength(s) of light as the laser.

References

1. K.C. Smith (ed.), The Science of Photobiology, Plenum Press, N.Y. (1977)
2. R.M. Tyrrell, Radiation Synergism and Antagonism, in "Photochemical and Photobiological Reviews, Vol. 3" (K.C. Smith, ed.), pp. 35-113, Plenum Press, N.Y. (1978).

Basic Photobiology and Open Problems

A.C. Giese

Department of Biological Sciences, Stanford University
Stanford, CA 94305, USA

1. Introduction

That light affects plants, animals, and humans has been known for a long
time, as enshrined in the myths concerning the sun in many ancient cultures
[1]. The Egyptian sungod Ammon Ra became the supreme deity in the pantheon
of gods; the very word radiation is said to stem from Ra. The sun god is
illustrated on Egyptian tomb decorations giving off light rays to plants,
animals and people, each ray ending in the symbol of life.

The supposed curative action of sunlight resulted in widespread use of
solaria by the ancient Greeks and Romans. The bleaching of pigments by sunlight was also familiar to the ancients. However, photochemical experiments
elucidating the way in which light acted on matter were begun only about two
hundred and fifty years ago [2].

I was asked to discuss basic photobiology as a background for the present
conference on Lasers in Photobiology and Photomedicine and to call attention
to open problems in photobiology to some of which laser techniques might be
applicable. Inasmuch as photobiology has its roots in photochemistry, I
have organized my subject matter with reference to photochemical principles,
which have to do with the absorption of light energy by molecules and the
way in which the energy is dissipated. It was early evident that not only
should the wavelength of light in photochemical experiments be known but
also that the amount of energy applied be determined. Quantitative studies
are equally important in photobiology [3,4].

2. Sunlight

The primary source of light on earth is the sun. Sunlight includes radiation over the entire span of the spectrum from ionizing radiation to radio-
waves [5]. However, most of the energy is in the span from the ultraviolet
(UV) to the infrared. About 5 percent of the energy is in the UV; about 35
percent is in the visible and the remainder, about 60 percent, is in the
infrared. Sunlight entering the atmosphere has a wider spread of wavelengths
than at the earth's surface, much of the short end of the UV spectrum being
absorbed by the ozone layer. The ozone layer, with a peak concentration at
about 25-30 km above sea level, is formed when oxygen absorbs the sun's very
short wavelength UV radiation.

Even with such attenuation from absorption by oxygen and ozone, with peaks
at 150 nm and 260 nm, respectively, enough short wavelength UV light comes
through the atmosphere to cause sunburn, skin aging and skin cancer in man,

and to kill and induce mutations in microorganisms. Most cells have developed repair systems enabling them to survive attack by such radiations [6].

The absorption spectrum of ozone resembles that of nucleic acids, the biochemicals in the cell most critical to cell survival. Attenuation of these damaging UV wavelengths makes the ozone layer a prime factor in the fitness of the environment for life on earth.

Sunlight is too variable during the day and with season to be useful as a source of radiation for most experimental purposes. The variability is partly a result of the changing angle of the sun with respect to the earth and the depth of atmosphere transversed, in part a result of atmospheric conditions, - for instance, the presence of fog, smog, rain and clouds. Daylength also changes with season and latitude. Consequently, most experiments are performed with artificial sources of light, amongst which lasers are useful where a small intense beam of light is effective, or where monochromatic light is desirable to cover relatively small surface areas. Lasers are available for infrared, visible and UV regions of the spectrum.

3. Grotthus-Draper Law

That light must be absorbed to provide energy for a photochemical or photobiological reaction now seems almost intuitive. However, a clear understanding of this relation was proposed by Grotthus only in 1818 and experimental support for it was provided by Draper in 1839 [2].

Molecules must have a chromophore (color-carrier) to absorb light. A chromophore is usually a series of alternate double bonds, among the prominent ones being ring structures, single, double, or multiple. The greater the number of rings attached to one another, the longer the wavelength of light absorbed. Dyes provide good examples of chromophores, absorbing in different parts of the spectrum. The chromophoric action of a given bond system may be shifted towards longer wavelengths by adding to the ring structures, groups known as auxochromes, for example, amino and hydroxyl groups.

Most organic molecules absorb UV light, especially the far UV, and at very short wavelengths almost all substances, even water (below 150 nm), become completely opaque to light. Ultraviolet absorption spectra are often distinctive enough to identify organic compounds.

Visible light (390-700 nm) is absorbed only slightly by unpigmented cells so that under the microscope the cells appear transparent, or translucent from scattering of light by particles in the cells. A little visible light is absorbed by all cells because of the presence of cytochromes and flavoproteins in catalytic amounts [7]. When a pigment is present some of the light is absorbed; that not absorbed is reflected, as for example the green color of chlorophyll in plants. The light energy absorbed by chlorophyll at the red and blue ends of the spectrum is partly stored as chemical energy in the products of photosynthesis.

Even unpigmented living cells absorb short wavelength UV light, some structures, for example, the chromosomes, are particularly prominent in photomicrographs taken with these rays. The core of a eukaryotic chromosome is a densely coiled deoxyribonucleic (DNA) molecule, and DNA absorbs short wavelength UV light strongly. Ribonucleic acid (RNA) is also present in the nuclear region outside the chromosome and in the cytoplasm. Both DNA and RNA absorb

UV light in a similar manner with a peak at 260 nm, and another peak in the very short wavelength range (Fig.1). Both DNA and RNA contain purine and pyrimidine bases, pentose sugars and phosphoric acid. The UV-absorbing moieties of the nucleic acid molecules are the organic bases.

About 15 to 20 percent of a cell consists of protein and proteins also absorb UV light with a peak at 280 nm and another at very short UV wavelengths (Fig.1). Such absorption is primarily a consequence of the presence in proteins of aromatic amino acid moieties, mainly tryptophane and tyrosine, which absorb at these wavelengths. The relative absorption of proteins and nucleic acids is quite different: a 0.02 percent solution of nucleic acid absorbs almost as much at its 260 nm peak as a 1 percent solution of protein absorbs at its 280 nm peak. For that reason chromosomes stand out clearly against the cytoplasm, although the latter also absorbs some of the same wavelengths of light. Protein absorption per unit weight of short wavelength UV light is less than that of nucleic acids because the proportion of aromatic light-absorbing amino acids in proteins is less than that of purine and pyrimidine bases in nucleic acids.

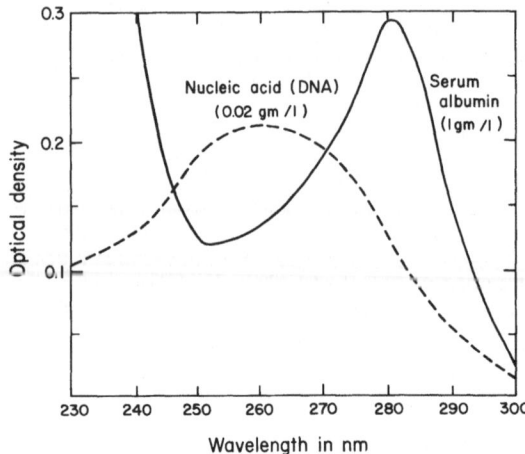

Fig.1 Absorption of UV light by nucleic acid (DNA) and protein (serum albumin). From A.C.Giese, 1979 Cell Physiology, 5th ed W.B. Saunders Co.,Philadelphia

Infrared light is absorbed by water. Cells composed of about 80 percent water absorb infrared light, but in natural light the amount absorbed is too slight to raise the temperature either locally in the cell or in the medium. However, if cells are exposed to a CO_2 laser beam they are killed and disrupted by its intense infrared light.

The effects of visible light are greatly exaggerated by the addition of a dye that complexes with molecules in the cell photosensitizing them, rendering them vulnerable to visible light. The dye, activated by absorbing visible light, alters cell organelles, usually by producing toxic compounds.

Dyes such as acridines and furocoumarins, intercalated among the DNA bases, transfer absorbed light energy that alters the DNA. Light-damaged DNA may not replicate, or it replicates altered bases, thereby producing a mutation. The mechanism of oxidative alteration of cells by photodynamic dyes is discussed in this volume by GIULIO JORI and the mechanism of furocoumarin action on DNA, by GIOVANNI RODIGHIERO.

Illumination of dyes localized in lysosomes damages the boundary membrane, releasing the hydrolytic enzymes from the lysosomes, much as UV light releases them from skin cells [6]. The enzymes digest the cellular organelles. Still other photodynamic dyes affect the cell membrane, altering the permeability of the cell.

Photosensitization of cells in human skin and other cells can be produced by natural plant photosensitizers, for example hypericins and furocoumarins; by photosensitizers in pathological or hereditary disorders of animals, for example, by some porphyrins; by some medications; and by pollutants in the air. Oxygen is required for photosensitization by hypericins and porphyrins, and by chlorophyll in carotenoid-deficient plant cells [8,9,10].

Photoallergies are also recognized in human skin. Chemicals, including some drugs, on or in the skin, are altered by exposure to light to become allergens that induce antibody formation. A second exposure to light then leads to a photoallergic reaction as the antibodies formed during the first exposure react with the allergen. The molecular biology of these photo-sensitizations is only partially elucidated [6].

In the presence of photosensitizers (for example, the furocoumarin 8-meth-oxypsoralen), psoriasis, an incredibly itchy and disfiguring overgrowth of human epidermal cells, can be controlled by illumination with long wavelength UV light. The white unpigmented spots on skin (vitiligo) are treated by application of similar photosensitizers and subsequent exposure to long wave-length UV, thereby inducing pigment formation [6]. Some tumors infused with photodynamic dyes and illuminated have been found to regress [12]. Neonatal jaundice (hyperbilirubinemia) is effectively resolved by treatment with blue-violet light alone [11,12]. The molecular basis of the responses in all the cases mentioned above is still uncertain.

When a conspicuous pigment is present in cells, it generally serves as the primary receptor for a photobiological reaction. For example, chloro-phyll is the light receptor for photosynthesis. When light has an effect on a colorless cell, the investigator attempts to ascertain the nature of the receptor by determining an action spectrum for the effect [13]. An action spectrum measures the relative efficiency of different wavelengths in pro-ducing a photobiological effect, for example the bactericidal action of UV light. The action spectrum is presumed to represent the absorption spectrum of the receptor, as predicted by the first law of photochemistry.

The action spectrum for killing of bacteria by UV light was demonstrated to be similar to absorption by nucleic acid [14], even though the absorption of light by the entire cell was found to be different from nucleic acid absorp-tion. The postulation of DNA as the receptor molecule for the damaging effects of UV light on cells was later verified by extracting the DNA and demonstrating changes in its structure, the most conspicuous of which is formation of pyrimidine dimers, especially thymine dimers. This topic is discussed in this volume by KENDRIC SMITH. Mutation by short wavelength UV light having an action spectrum similar to nucleic acid absorption is con-sidered to be the result of altered DNA in the cells surviving irradiation [15]. The action spectrum for induction of chromosome aberrations also resembles nucleic acid absorption [16].

Changes in properties of a virus or cell by UV light may be caused by action on different receptors in the cell as shown by action spectra. For example, the action spectrum for the loss in infectivity of the influenza virus resembles the absorption of UV light by nucleic acid, whereas the action spectrum for inactivation of its red blood cell agglutinating property [17] resembles absorption of UV light by a protein (Fig.2). In similar manner, the action spectrum for retardation of division of paramecia resembles absorption by nucleic acid, while that for immobilization resembles absorption by proteins. Still other types of UV action spectra suggesting receptors other than nucleic acids or proteins have been described [18]. Action spectra for photosensitization by dyes resemble the absorption spectra for the dyes, as anticipated.

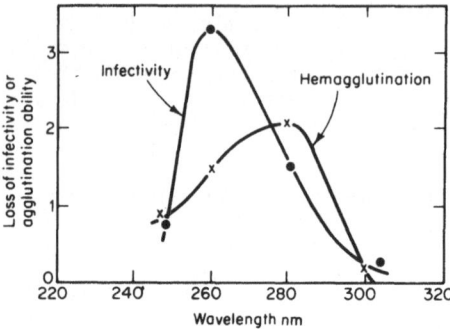

Fig.2 Action spectrum for inactivation of virus and loss of agglutinating activity on red blood cells. From A.D. McClaren and D. Shugar, 1964 Photochemistry of Proteins and Acids. Pergamon Press, New York

Under some conditions an action spectrum may be misleading. For example, the erythema (reddening) action spectrum of the human skin shows a peak at 297 nm with low values to either side. This does not resemble the absorption spectrum of any cell constituent; it is rather the result of skin structure. The incident light must pass through a layer of dead cells that filter the light; the shortest wavelength with damaging action that reaches the living cells in the prickle cell layer is 297 nm [13].

4. Bunsen-Roscoe Reciprocity Law

The Bunsen-Roscoe reciprocity law states that as long as the product of the intensity of the light and the time of exposure (i.e., the dose) is the same, the photochemical effect will be the same. That is, for the same dose, low intensity light is as effective as high. A small degree of reciprocity failure is found in some photochemical reactions. In biological systems, on the other hand, for the same dose, cells are less affected when treated with low than with high intensity short wavelength UV light because repair reverses some of the damage. Several types of repair of radiation-induced as well as chemical damage have been found and are described by KENDRIC SMITH in this volume. Repair may be an important feature of many aspects of life processes.

Meriting attention and further study are some contrary results for retardation of division by short wavelength UV light upon ciliates. In this case the delay of division by a given dose of UV light is greater for low than for high intensity light. Flashing light, permitting an interval of time between applications of UV light, is even more effective than continuous light, suggesting secondary thermal reactions following photochemical action [19].

5. The Lambert-Beer Law

The Lambert-Beer Law of photochemistry states that light entering an absorbing solution is attenuated exponentially. Similarly, UV light entering the cytoplasm of a cell is attenuated exponentially. When the cell is small, as in a bacterium, the UV light intensity is attenuated only slightly before the DNA chromosome is reached. In an animal cell, for example a sea urchin egg or a large protozoan, the nucleus is screened by a thick layer of cytoplasm. In a plant cell the cell wall also attenuates short wavelength UV light. The response of a eukaryotic cell to UV irradiation may therefore differ considerably from that in a prokaryotic cell.

Furthermore, a eukaryotic cell is organized into cell organelles, each of which performs a function unique to it [6]. For example, the mitochondria are the seat of energy-liberating reactions forming ATP, which serves as the currency of the cell. Many syntheses occur on the membranes of the endoplasmic reticulum-Golgi body complex from which the lysosomes and various other products take origin. In green plant cells photosynthesis occurs in the chloroplasts. Proteins are synthesized on ribosomes attached to the surface of the endoplasmic reticulum. Therefore, in a eukaryotic cell, UV radiation damages not only the DNA in the cell nucleus but also disrupts the normal functions of cellular membrane systems; even if the cells survive irradiation they function abnormally. This is shown by changes in movement, permeability and excitability, functions more readily measured on eukaryotic than on prokaryotic cells. To induce loss in viability and mutation in eukaryotic cells requires larger doses of radiation than in prokaryotic cells because of the lesser accessibility of DNA, upon which these depend, to incoming light. When the protective cytoplasm is diminished in thickness, as in starving paramecia, the sensitivity of the cell to UV light is increased [20].

As a consequence of the greater screening action of the cytoplasm present in a sea urchin egg than on a sea urchin sperm, the susceptibility of sperm and egg to short wavelength UV light differs by several orders of magnitude. To produce an equal degree of retardation of cell division only one millionth the dose need be administered to sperm used to fertilize unirradiated eggs as to eggs fertilized with unirradiated sperm [21]. A sperm is essentially a large nucleus with a thin coat of cytoplasm, whereas the egg has a thick layer of cytoplasm, loaded with nutrient for development of an embryo, surrounding the nucleus [21].

That nuclear damage, or more specifically DNA damage, should be so important in action of UV radiation on cells is not unexpected considering the multiplicity of various UV light receptors in the cell and their function. DNA is present in most cell chromosomes only in duplicate, whereas RNA and proteins are present in multiple copies. Consequently, destruction of some of these molecules is not as damaging to cell viability as damage to DNA. Furthermore, all of these molecules can be resynthesized if an undamaged DNA template is available.

Laser techniques are particularly valuable for studies on the highly compartmentalized eukaryotic cell because the beam of light can be focused on a particular organelle or some fraction of it, as shown by the studies by MICHAEL BERNS [22]. The function of the cell components may therefore be ascertained by the change in cell behavior when a particular cell constituent is irradiated. Numerous earlier studies on effects of microbeams on cells have been summarized [23].

Visible light is able to reach cell components in photosensitized cells because the cell is essentially transparent to visible light. Attenuation of light is therefore less of a problem in studies with laser beams treating cells sensitized by dyes than with short wavelength UV light.

6. Photoequivalence and the Mechanism of Photobiological Reactions

The Einstein-Stark law of photochemical equivalence states that for every quantum absorbed, an activated particle (atom, molecule, free radical) is formed. This is called the primary reaction. It is followed by secondary reaction(s). The biological system is no exception to the rule.

Absorption of a quantum of light by a receptor molecule leads to its excitation. Excitation may be dissipated quickly in fluorescence at longer wavelengths or more slowly by phosphorescence, or the absorbed energy may be degraded to heat. On the other hand, the excited receptor may transfer its energy to other molecules, as in photosensitization, or react to form photo-products in secondary reactions. For example, highly reactive species called free radicals may be produced. These can be detected and followed by electron spin resonance and much has been learned about photochemical reactions using this technique. The free radical may react with oxygen to form an oxidized cell constituent. In photosensitization, highly reactive singlet oxygen may be produced. This in turn may oxidize cellular constituents [10].

Oxygen is therefore important in many photobiological reactions. Oxygen is often required for bioluminescence - the production of light by some organisms [24]. Oxygen is always present in aerobic organisms and is produced by photosynthetic plants. Interestingly, chlorophyll is a photosensitizer, therefore a photosynthetic plant must protect its photosynthetic apparatus from photooxidation. This is accomplished by the presence of carotenoids with conjugate double bond systems that apparently, in a manner yet unknown, protect the cell constituents from the photosensitizing action of chlorophyll. Mutants without such protection are subject to photosensitized damage [8].

Oxygen is not necessary for photosensitization by furocoumarins, and it may even interfere with some of their activities by inactivation of light-excited molecules [25]. Oxygen is not required for action of short wavelength UV radiation, the primary process depending upon photochemical alteration of DNA or other compounds in the cell. However, oxygen may be important in secondary reactions. When ozone is formed by very short UV light it damages the cells [26]. Under some conditions hydrogen peroxide or organic peroxides may be formed by UV radiation [18].

7. Open Problems in Photobiology

Besides the effects of UV radiation on cell activities described above and photosensitization, photobiology treats other problems that may be benefited by laser application. Among these problems are vision, photomovement and other photoresponses, photosynthesis, photic nitrogen fixation, photomorphogenesis, circadian rhythms, photoactivation of enzymes, visible light effects, photomedicine, photoantagonism, photosynergism and bioluminescence. Photomedicine has already been referred to above in some instances and is amply covered in other reports in this volume and in the literature [12].

The receptor molecules are known for vision in vertebrates (rhodopsin and iodopsin, for dim and bright light vision, respectively), photosynthesis (chlorophyll and accessory pigments), and photomorphogenesis (phytochrome). However, the receptor molecules are not known for photoresponses in many primitive organisms and for light synchronization of circadian rhythms. Action spectra might help define the receptor molecules; for this, monochromatic tunable lasers might be effective. Considerable information on the molecular biology of bioluminescence is at present available. A brief summary of the state of the problems in each of these fields is outlined below [27].

7.1 Photoresponses

Photoresponses are an important cue for orientation of most organisms. Bacteria, single-celled zoospores and gametes of lower plants, protozoans, and animals without eyes respond to light in well-characterized patterns [28]. However, photobiological mechanisms have been analyzed at the molecular level primarily in vertebrate eyes. Duplex photoreceptors, present in man and a number of other vertebrates, make possible the detection of light over a 10^{10} intensity range; the rods serve for dim colorless vision, the cones, for bright light color vision. The outer segment of the rod contains numerous disk-shaped membranes that form a closely packed stack of disks, each disk loaded with the visual pigment, rhodopsin. Rhodopsin consists of the protein, opsin, conjugated with retinal, a cis-isomer of vitamin A aldehyde. Absorption of light leads to a change in retinal to form all trans-retinal.

The conformational change in the rhodopsin molecule induced by light triggers hyperpolarization of the cell via a transmitter, most likely released calcium ions. The calcium ions alter the permeability of the cell surface causing leakage of ions and hyperpolarization of the membrane. Connecting neurons activated by the rod send trains of impulses to the brain. Similar events occur in the cone that contains the pigment iodopsin [28a]. Much less is known about the photoreceptor process in invertebrates. Tunable lasers might serve for determination of action spectra in such cases.

Light is also important in the orientation of plants throughout life. Analysis of the photoreceptor process in plants is beset with difficulties. In higher plants the major photoreceptive pigment is probably a flavoprotein, as deduced from action spectra and corroborative studies, not a carotenoid as was inferred from studies on animals. Some pigment(s) other than flavoproteins may also play a minor role. Absorption of light by the flavoprotein pigment results in a differential distribution of the hormone, auxin, which in the stem accumulates away from the light. This leads to growth or elongation of the side away from the light, resulting in bending of the stem towards the light. Leaves orient with respect to light in a similar manner, and in "compass" plants, they reorient throughout the day. Some flowers (for example, the sunflower) face the sun throughout the day. But how, at the molecular level, light causes the redistribution of auxin is still largely speculative [28].

7.2 Photosynthesis

In photosynthesis light energy absorbed by chlorophyll is transduced to chemical energy by the production of high energy phosphates (photophosphorylation). Adenosine diphosphate (ADP) is photophosphorylated to adenosine triphosphate (ATP), which furnishes the energy for cellular work. Light also

provides the energy for reduction of coenzymes that participate in carbon
dioxide fixation. Only a fraction of the chlorophyll is present at the
reaction center at which the energy transduction occurs, perhaps a single
molecule per center, the remaining chlorophyll and accessory pigments act
as an antenna to absorb light energy and funnel it to the reaction center.
Determination of the nature of the reaction centers and manner in which they
perform their role in energy transfer are key problems in photosynthesis.
Another unresolved problem is the mechanism of oxygen evolution [29].

For studies on the fast reactions recorded spectrophotometrically, short
pulse laser radiation has been especially useful in analysis of photosynthesis. The data have recently been summarized [30].

7.3 Photic Nitrogen Fixation and Production of Hydrogen

Less evident, but maybe equally important, is the use of light energy for
nitrogen fixation. The fixed nitrogen supply of bodies of water (and soil)
limits growth of plants. While some textbooks lead one to believe that the
nitrogen cycle is a function of various bacteria, the blue-green algae also
play an important role, especially in fresh and salt waters. Nitrogen fixation is an energy-demanding process and soils and natural waters seldom have
much organic matter to provide energy to extensive nitrogen fixation by non-
photosynthetic bacteria, but blue-green algae can use the energy of sunlight
for this purpose. Therefore, when present, blue-green algae are important
contributors to the fixed nitrogen supply of the biosphere. The mechanism by
which they use the light energy to reduce atmospheric nitrogen to ammonia is
only partially understood [31].

One problem that might be elucidated with laser beams is the function of
heterocysts in blue-green algae. Heterocysts are large, weakly colored,
thick-walled cells, present in many filamentous species, and are thought to
serve for nitrogen fixation in the species possessing them. With conventional
methods it is difficult to illuminate single cells in a chain to determine
whether they perform photosynthesis or nitrogen fixation or both, something
quite feasible with laser beams. Selective illumination of either green cells
or heterocysts and determination of photosynthesis and nitrogen fixation with
tracers would permit resolution of the problem.

Photosynthetic cells, including the blue-green algae, can be adapted to use
light energy to liberate hydrogen from water. Should this be possible on a
practical scale, it may provide an alternate supply of fuel provided we learn
to control the process. Control may be possible once the mechanism of hydrogen liberation is resolved [31].

7.4 Circadian Rhythms

Light is a major cue for synchronizing daily rhythms in cells with the 24-hour
day-night cycle of the earth's environment. Most of the rhythms in plants,
fungi and animals (invertebrate and vertebrate) are circadian, that is, almost
but not quite 24 hours when free running in the dark or under other constant
conditions. For example, many plants open their petals or move their leaves
at a specific time of day. Plants kept in continuous dim light continue to
show the same rhythms but the rhytnms are now circadian. Normal day and
night exposure synchronizes the rhythms to exactly 24 hours. Bees that depend
upon nectar and pollen from plants know by their own biological clocks when
to come out to feed upon the species of plants of their liking. This rhythm

is also circadian under constant conditions. Squirrels have similar rhythms, as does the human. These rhythms are inherent (endogenous) and are timed by a biological clock, the molecular nature of which has not been determined [32].

In spite of the universality of circadian rhythms in nucleated (eukaryotic) cells, no universal chromophore has apparently evolved as a receptor serving to synchronize circadian rhythms to the 24 h day. Lower plants use blue-absorbing pigments like flavoproteins and possibly carotenoids; higher plants depend upon chlorophyll and phytochrome as membrane-bound photoreceptive pigments for the purpose. Much also remains to be learned about the location of these pigments. Even less is known about photoreceptors synchronizing circadian rhythms in animals, mainly because most behavioral studies have been performed with white light. Only in the fruit fly (Drosophila) and the pink bollworm (Pectinophora) is the photoreceptor known to be a flavoprotein, present in brain-associated extraretinal receptors. Action spectrum studies are therefore needed to establish the nature of chromophores in other animals. In many invertebrates extra-retinal photoreceptors are the site of the chromophores, as they are in lower vertebrates but in adult mammals the eyes serve this function [33].

In view of the fact that the dose of light required to synchronize circadian rhythms with the 24 h day is very low, action spectra determined with wide band filters should be redetermined with monochromatic light. Stray light from filters might well have led to erroneous results.

7.5 Photoactivation of Enzymes

Although reports early in the present century indicated activation of some enzymes by light, only recently have experiments suggested diverse mechanisms by which light activation occurs. Because some enzymes so activated may initiate a series of successive enzyme activations by interaction with allosteric control enzymes, light activation may have a significant effect on metabolism of cells. It is to be expected that as other enzymes are studied much interesting information on how light affects metabolism, previously unsuspected, will be revealed [34,35].

7.6 Visible Light Effects

Because cells appear to be transparent, visible light is presumed to be innocuous. However, if visible light is absorbed by critical molecules in a cell, it can be as effective as UV light in producing photobiological effects. A few effects of visible light only slightly explored are: retardation of cell growth in culture, induction of chromosome aberrations and slowing of DNA synthesis in tissue culture, mutagenesis, and killing in repair-deficient bacteria [7]. The chromophores for these effects are thought to be primarily cytochromes and riboflavin enzymes [7]; others are unidentified.

Except for its effect on visual acuity, we do not know how lighting affects human behavior and well-being. Optimal lighting may improve human performance in home and industry. Environmental photobiology is only developing as we begin to appreciate that the sun is a major component of the natural environment on earth [34].

We do not know whether coherent light emitted by lasers has a different effect on biological systems than non-coherent light emitted by hot bodies. Considering the large number of photobiological problems studied with non-coherent light, we realize the need for further exploration of this aspect.

7.7 Photoantagonism

Sometimes the effect of exposure to one part of the spectrum producing a biological effect is negated by simultaneous or subsequent exposure to another part of the spectrum. One of the most interesting examples is photoreversal of damage to cellular DNA caused by short wavelength UV light by blue-violet and long wavelength UV light. As much as 80 percent of the damage from short wavelength UV light to DNA may be photoreversed in this manner; the various other types of damage to DNA are not [36]. The molecular mechanism for photoreversal is discussed in this volume by KENDRIC SMITH.

Antagonism occurs between the effects of red (650 nm) and far red (770 nm) light upon the photoregulatory pigment, phytochrome, present in plants. Phytochrome is a bluish pigment consisting of an open tetrapyrrol attached to a protein. It regulates photomorphogenesis - that is, growth and developmental processes in plants, including flowering. The far red-absorbing form of the pigment is considered to be the active form that sets in action other enzymes (perhaps 24 in all) that regulate growth and form in plants. Absorption of red light induces a conformational change converting the inactive into the active form of the enzyme [37,38]. In both cases cited above discovery of antagonistic action gave considerable insight into the mechanism of action of light on the photoreactive system.

7.8 Photosynergism

When irradiation with two monochromatic wavelengths of light has more than an additive effect on a biological system than the sum of each of the wavelength singly applied, the effect is said to be synergistic. For instance, in photosynthesis, if x is the rate when one wavelength alone is applied, for example, 700 nm, and y is the rate when a shorter wavelength alone is applied, for example, 650 nm, then enhancement (synergism) occurs if z, the rate when both wavelengths are applied, is greater than the sum x + y [39]. This discovery had a profound effect on concepts of the mechanism of photosynthesis.

Evidence for synergism has been presented for a wide variety of combined radiations. For example, x-rays and UV light, x-rays and red light, x-rays and infra-red light, x-rays and microwaves, x-rays and ultrasound, short and long wavelength UV rays [40]. Because preconditioning tumors by irradiation with less damaging radiation is favorable for treatment of tumors with ionizing radiation, synergism offers great advantages for treatment of tumors in humans.

Synergism depends upon the genotype of the cells. For example, a dose of 365 nm UV radiation that inactivates 30 percent of the E. coli population strongly sensitizes the surviving cells to subsequent 254 nm UV radiation. The effect varies in different strains of bacteria, being more prominent in repair-deficient strains. A repair-competent strain of E. coli (K12 AB1157) showed a lesser degree of synergism than a repair deficient strain, and wild type cells showed even less. Dependence of synergism on various factors is discussed in a recent review on synergism and antagonism [40].

7.9 Bioluminescence

Photobiology also deals with the production of light by organisms. Some cells produce light of high efficiency for species with light receptors. Bioluminescence in its simplest form results from the oxidation of a substrate, luciferin, by oxygen in the presence of an enzyme, luciferase, as in the Japanese "water firefly" Cypridina, but a variety of systems has been described. Much progress has been made in the molecular biology of the reactions [41]. Recent studies have focused on energy transfer and fluorescent proteins that in some species serve as energy acceptors for the oxidized excited chromophore and reinforce the luminescence [24]. Inasmuch as this conference is concerned with effects of laser radiation on biological systems, further consideration of the topic is irrelevant.

Conclusion

Photobiology deals with the effect of the span of light, from the infrared to the ultraviolet, which comes from the sun. Without sunlight there could be no life on earth because the sun's energy powers life activities. However, because light comes from the sun it is not necessarily beneficial; some effects of sunlight are distinctly damaging. Our sun has two faces - one good and one bad like Janus the Roman sun god [6]. Some damaging effects of one span of sunlight are ameliorated by another span (antagonistic action), for example, damage by short wavelength UV light is photoreversed by visible and long wavelength UV light. Life accommodates to other damaging action of light by behavior - avoiding the light, or by protective devices such as pigments and opaque screens of exoskeletons that prevent the incident light from reaching inside the cell. Furthermore, cells have also developed very effective dark repair mechanisms for reversing the damage by light entering the cell to cellular DNA. DNA is the most critical biochemical in the cell because of its uniqueness, it generally is present in the cell only in duplicate whereas the other biochemicals, such as RNA and proteins, are present in multiple.

Study of the responses of life to light has led to considerable insight into mechanisms by which cells operate and as to the pathology of cells underlying disease. Lasers might be of considerable value as probes in exploring physiological processes in cells. In some cases this can be done by further investigation of photobiological problems, as suggested in the discussion of basic photobiology; in others, lasers may serve as surgical tools for selective destruction of cell organelles. Comparison of functioning of cells after destruction of an organelle may lead to a better understanding of the normal function of the organelle.

References

1. W.J.Olcott: Sun Lore of All Ages (Vantage Press, New York 1914)

2. A.C.Giese: In Photophysiology Vol.1 ed. by A.C.Giese (Academic Press, New York and London 1964) pp.1-18

3. J.Jagger: Introduction to Research in Photobiology (Prentice Hall, Englewood Cliffs, New Jersey 1967)

4. K.C.Smith, ed.: The Science of Photobiology (Plenum Press, New York 1977)

5. E.N.Parker: Scient. Amer. (Sept) $\underline{223}$, 43-50 (1975)

6. A.C.Giese: <u>Living With Our Sun's Ultraviolet Rays</u> (Plenum Press, New York 1976)

7. B.Epel: In <u>Photophysiology</u> Vol. 8 ed. by A.C.Giese (Academic Press, New York and London 1973) pp. 209-229

8. N.I.Krinsky: In <u>Photophysiology</u> Vol. 3 ed. by A.C.Giese (Academic Press, New York and London 1968) pp. 123-195

9. A.C.Giese: In <u>Photophysiology</u> Vol. 6, ed. by A.C.Giese (Academic Press, New York and London 1971) pp. 77-129

10. J.Spikes: In <u>The Science of Photobiology</u> ed. by K.C.Smith (Plenum Press, New York 1977) pp. 87-112

11. T.R.C.Sisson: <u>Photochem. Photobiol. Rev.</u> Vol. 1, ed. by K.C.Smith (Plenum Press, New York 1976) pp. 241-268

12. J.H.Epstein: In <u>The Science of Photobiology</u> ed. by K.C.Smith (Plenum Press, New York 1977) pp. 125-207

13. A.C.Giese: Photochem.Photobiol. $\underline{8}$, 127-146 (1968)

14. F.L.Gates: Science $\underline{68}$, 479 (1929)

15. K.C.Smith: In <u>The Science of Photobiology</u> ed. by K.C.Smith (Plenum Press, New York 1977) pp. 113-142

16. J.S.Kirby-Smith and D.L.Craig: Genetics $\underline{42}$, 123-175 (1957)

17. I.Tamm and D.J.Fluke: J. Bact. $\underline{59}$, 35-113 (1950)

18. A.C.Giese: Physiol. Zool. $\underline{18}$, 223-250 (1945)

19. A.C.Giese, D.C.Shepard, J.Bennett, A. Farmanfarmaian, and C.L.Brandt: J. Gen. Physiol. $\underline{40}$, 311-325 (1956)

20. A.C.Giese: J. Cell. Comp. Physiol. $\underline{26}$, 43-55 (1945)

21. A.C.Giese: Biol. Bull. $\underline{91}$, 81-87 (1946)

22. M.W.Berns: <u>Photochem. Photobiol. Rev.</u> Vol. 3, ed. by K.C.Smith (Plenum Press, New York 1978) pp. 1-37

23. M.W.Berns: <u>Biological Microirradiation, Classical and Laser Studies.</u> Prentice-Hall, Englewood Cliffs, New Jersey (1974)

24. W.W.Ward: In <u>Photochem. Photobiol. Rev.</u> Vol. 4 ed. by K.C.Smith (Plenum Press, New York 1979) pp. 1-57

25. L.Musajo and G. Rodighiero: In <u>Photophysiology</u> Vol. 7, ed. by A.C. Giese (Academic Press, New York and London 1972) pp. 115-147

26. A.C.Giese and E.Christensen: Physiol. Zool. $\underline{27}$, 101-115 (1965)

27. A.C.Giese: Bioscience <u>29</u>, 353-357 (1979)

28. W.G.Hand: In <u>The Science of Photobiology</u> ed. by K.C.Smith (Plenum Press, New York 1977) pp. 313-328

28a. E.A.Dratz: In <u>The Science of Photobiology</u> ed. by K.C.Smith (Plenum Press, New York 1977) pp. 241-279

29. D.C.Fork: In <u>The Science of Photobiology</u> ed. by K.C.Smith (Plenum Press, New York 1977) pp. 329-369

30. Govindjee and P.A.Jursinic: <u>Photochem. Photobiol. Rev.</u> Vol. 4, ed. by K.C.Smith (Plenum Press, New York 1979) pp. 125-205

31. J.Postgate: <u>Nitrogen Fixation</u>, Edward Arnold, London (1978)

32. B.Sweeney: In <u>The Science of Photobiology</u> ed. by K.C.Smith (Plenum Press, New York 1977) pp. 209-240

33. H.Ninnemann: <u>Photochem. Photobiol. Rev.</u> Vol. 4 ed. by K.C.Smith (Plenum Press, New York 1979) pp. 207-266

34. K.C.Smith: In <u>The Science of Photobiology</u> ed. by K.C.Smith (Plenum Press, New York 1977) pp. 397-417

35. D.H.Hug: <u>Photochem. Photobiol. Rev.</u> Vol. 3 ed. by K.C.Smith (Plenum Press, New York 1978) pp. 1-33

36. R.O.Rahn: <u>Photochem. Photobiol. Rev.</u> Vol. 4 ed. by K.C.Smith (Plenum Press, New York 1979) pp. 267-330

37. W.Shropshire Jr.: In <u>The Science of Photobiology</u> ed. by K.C.Smith (Plenum Press, New York 1977) pp. 281-310

38. L.H.Pratt: <u>Photochem. Photobiol. Rev.</u> Vol. 4 ed. by K.C.Smith (Plenum Press, New York 1979) pp. 59-124

39. R.K.Clayton: <u>Light and Living Matter</u>. 2 Vol. (McGraw-Hill, New York 1970/71)

40. R.M.Tyrrell: In <u>Photochem. Photobiol. Rev.</u> Vol. 3, ed. by K.C.Smith (Plenum Press, New York 1978) pp. 35-113

41. J.Lee: In <u>The Science of Photobiology</u> ed. by K.C.Smith (Plenum Press, New York 1977) pp. 371-395

42. R.B.Webb: <u>Photochem. Photobiol. Rev.</u> 2, ed. by K.C.Smith (Plenum Press, New York 1977) pp. 169-261

Photobiology of Ultraviolet Radiation

K.C. Smith

Department of Radiology, Stanford University School of Medicine
Stanford, CA 94305, USA

SUMMARY

Since enzymes respond to specific substrates, and since radiation produces
many different types of alterations in the structure of deoxyribonucleic acid
(DNA), it is not surprising that enzymatic pathways for the repair of damaged
DNA are numerous and complex. Most of these enzymatic pathways repair DNA
accurately, but some are inaccurate and therefore produce mutations. The
production and repair of DNA damage, and its role in mutagenesis are discussed.

INTRODUCTION

The sensitivity of a cell to ultraviolet (UV) radiation-produced lethality
depends upon the intrinsic sensitivity of its deoxyribonucleic acid (DNA) to
undergo photochemical alteration, and on the ability of the cell to accurately
repair such damage when it is produced.

Cells have evolved a complex series of biochemical pathways for maintaining the integrity of the structure of their DNA. Three major types of repair
are known: photoreactivation (utilizing a repair enzyme that requires light
energy for it to function), and two repair systems that do not require light
for their function; excision repair and postreplication repair. These last
two systems are each composed of several independent biochemical pathways.
The enzymes for most of these repair pathways are always present in the cell,
while some are only synthesized subsequent to the radiation insult. Most
of these repair pathways are accurate, but some are inaccurate and thereby
produce mutations. A mutation is believed to be the first step leading to
a cancer.

This report will describe the radiation-produced changes that can occur
in DNA, the multiple mechanisms for the repair of this damage, and the role
of DNA repair in mutagenesis.

RADIATION CHEMISTRY OF DNA

Radiation or chemical damage to DNA can be divided into two categories,
those that alter the purine and pyrimidine bases, and those that alter the
sugar-phosphate polymer and produce a break in the DNA chain. Excellent
reviews on the photochemistry (1) and radiation chemistry (2) of DNA have
appeared.

Base Damage. Unimolecular base damage is produced by such reactions as the addition of H·, OH· and HO$_2$· radicals to the purine and pyrimidine rings (Fig.1B,D,E), or the oxidation of a side chain (Fig.1C). These types of reactions are most efficiently produced by ionizing radiation, however, 254 nm-radiation is very efficient in producing pyrimidine hydrates (Fig.1B), and 313-nm radiation can produce the thymine glycol (Fig.1E) (3).

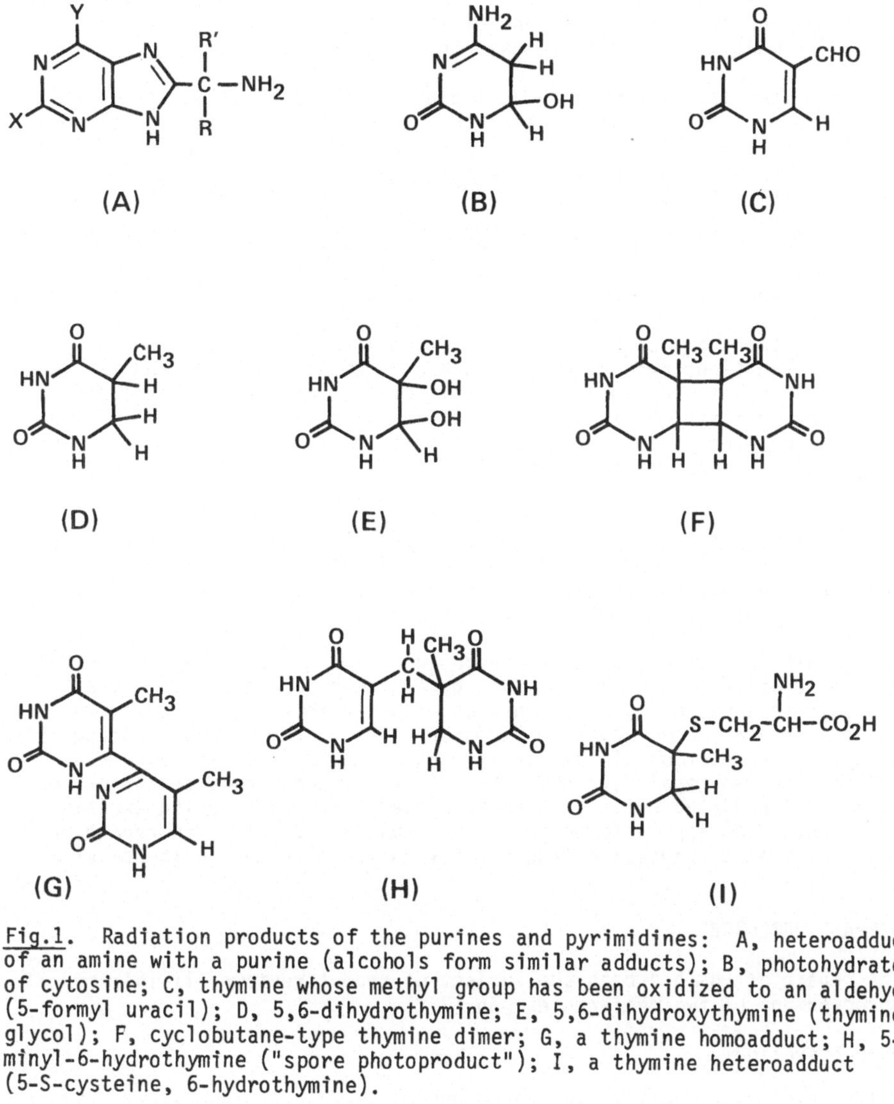

(A)   (B)   (C)

(D)   (E)   (F)

(G)   (H)   (I)

Fig.1. Radiation products of the purines and pyrimidines: A, heteroadduct of an amine with a purine (alcohols form similar adducts); B, photohydrate of cytosine; C, thymine whose methyl group has been oxidized to an aldehyde (5-formyl uracil); D, 5,6-dihydrothymine; E, 5,6-dihydroxythymine (thymine glycol); F, cyclobutane-type thymine dimer; G, a thymine homoadduct; H, 5-thyminyl-6-hydrothymine ("spore photoproduct"); I, a thymine heteroadduct (5-S-cysteine, 6-hydrothymine).

Under normal experimental conditions (i.e., wet cells at room temperature), the photoproduct that is produced in DNA in highest yield by 254 nm-radiation is the cyclobutane-type pyrimidine dimer (Fig. 1F); formed by the linkage of two pyrimidine residues that are adjacent in the same strand of DNA.

41

Recently, the cyclobutane-type thymine dimer was also isolated from cells exposed to ionizing radiation (4).

Other types of photochemical homoaddition reactions between pyrimidines are also known (e.g., Fig. 1G, H). In general, these types of products are produced in much lower yield than are the cyclobutane-type dimers, but this should not be interpreted to mean that they are therefore of less biological importance. The chemical nature and biological importance of nondimer damage produced in DNA by UV-irradiation has been reviewed (5).

Although the possible number of different types of unimolecular and bi-molecular homoaddition reactions that the purines and pyrimidines can under-go is limited, the number of possible reactions between the nucleic acid bases and other types of compounds (heteroaddition reactions) is almost un-limited. This is because cellular DNA is in intimate contact with proteins, lipids, carbohydrates, and numerous small molecular weight compounds, mol-ecules that can form heteroaddition products with the DNA.

A large number of different types of compounds have been shown to combine with the nucleic acid bases during irradiation (e.g., Fig.1A,I). Certain of these reactions are produced by the direct absorption of UV radiation, while other reactions require near-UV and/or visible radiation and the pres-ence of photosensitizing molecules. Still other bimolecular heteroaddition reactions are mediated by ionizing radiation. The chemical nature of DNA heteroaddition reactions and their importance in aging, carcinogenesis and radiation biology have been reviewed (6).

Strand Breaks. DNA-single-strand breaks are produced in low yield by the direct photochemical action of UV radiation (254 nm), but are produced in high yield by gamma and X radiations, and in vitro by photochemical re-actions mediated by near-UV radiation plus certain photosensitizers (7), and in vivo by near-UV radiation alone (8).

DNA double-strand breaks can be produced by two closely spaced chemical events on opposite strands of the DNA, by a single radiation event that breaks both strands, or by a combination of a single radiation event that breaks one strand and an enzymatic excision event on the opposite strand (9). The enzymatic production of DNA single-strand breaks is a necessary step in most of the DNA repair systems. DNA double-strand breaks are also pro-duced in UV-irradiated cells as a consequence of two overlapping excision repair events, and appear to be the major cause of lethality in repair pro-ficient cells at UV radiation fluences beyond the shoulder on the survival curve (10).

DNA REPAIR PROCESSES

The repair of DNA has been the subject of recent reviews (11-15), and of the proceedings of two conferences (16,17).

Photoreactivation

Cells that have been inactivated by UV irradiation can be reactivated by a second irradiation with visible light. The single enzyme responsible for this process has been isolated, and appears to act only on cyclobutane-type pyrimidine dimers (Fig.1F). The photoreactivating enzyme combines with a pyrimidine dimer in DNA, and when this enzyme-substrate complex is exposed

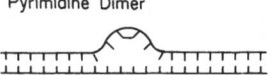

Pyrimidine Dimer

Enzyme-DNA Complex

Absorption of Light and Repair

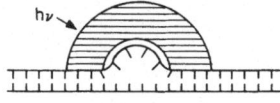

Release of Enzyme

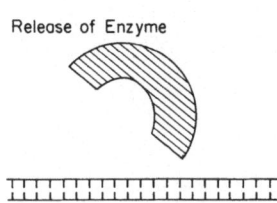

Fig.2. A general model for photoreactivation. The enzyme combines with cyclobutane-type pyrimidine dimers in UV-irradiated DNA to form an enzyme-substrate complex. The absorption of light between 320-410 nm activates the complex, the pyrimidine dimers are converted to monomeric pyrimidines, and the enzyme is released.

to visible light at about 400 nm, the dimer is converted to monomeric pyrimidines, and the enzyme is released. This is shown schematically in Fig. 2.

The extensive reversal of the lethal and mutagenic effects of UV radiation by photoreactivation has been taken as proof of the involvement of pyrimidine dimers in these processes. The same enzyme that is responsible for photoreactivation afer UV irradiation is also responsible for the small amount of photoreactivation seen under special conditions after γ-irradiation (4).

Excision Repair

The general principle of this type of repair is that the damage is recognized, cut out and replaced with undamaged material.

Nucleotide Excision. In this repair system a section of the DNA containing the damaged nucleotides is cut out and replaced with undamaged nucleotides. The scheme shown in Fig.3 is the general mechanism for the major pathway of excision repair, the one that can function even when DNA replication is inhibited (e.g., when cells are in buffer). However, there is also a minor pathway of excision repair that can only proceed when the cells are in complete growth medium. The growth medium-independent pathway appears to be error free, and to make short patches of repair replication, while the growth medium-dependent pathway makes long patches and appears to be error-prone, i.e., mutagenic.

43

INCISION

REPAIR REPLICATION

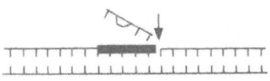

EXCISION

REJOINING

Fig.3. A general model for the major pathway of excision repair. An enzyme recognizes the lesion, shown here as a cyclobutane-type pyrimidine dimer, and makes an incision in the DNA strand. Repair replication (heavy line) commences using the opposite strand of DNA as the template. Finally, the damaged section of the DNA is excised, and the break in the DNA strand is sealed.

Base Excision. In this repair system a damaged base is recognized and removed from DNA without breaking the strand by a class of specific enzymes called N-glycosylases (i.e., they cleave the N-glycosylic bond of a nucleotide). Specific N-glycosylases that recognize uracil, hypoxanthine, 3-methyl adenine and 0^6-methyl guanine have been isolated. In principle, a missing base can be reinserted using the complementary base on the opposite strand as a template. Such "insertase" activity has been observed in extracts from human fibroblasts (18).

A more general system for the repair of these apurinic or apyrimidinic (AP) sites is excision repair initiated by an AP endonuclease. The AP endonuclease makes a chain break at an AP site, and then excision events similar to those shown in Fig. 3 are expected to occur.

Base excision repair is a relatively recent discovery (reviewed in 11,15, 19). It is expected that this repair system will prove to be especially important to cells that have been exposed to ionizing radiation or alkylating agents.

Postreplication Repair

The DNA that is synthesized shortly after UV irradiation in bacteria or mammalian cells has discontinuities when assayed in alkaline sucrose gradients. (In excision deficient cells of E. coli, the mean length of these newly synthesized daughter-strands of DNA approximates the average distance between pyrimidine dimers in the parental strands.) With further incubation of the cells, these discontinuities disappear, and the DNA approximates the molecular size of that from unirradiated cells. These resulults suggested the postreplication repair system shown schematically in Fig.4. DNA replication proceeds past the lesions in the parental strands, leaving gaps in the

44

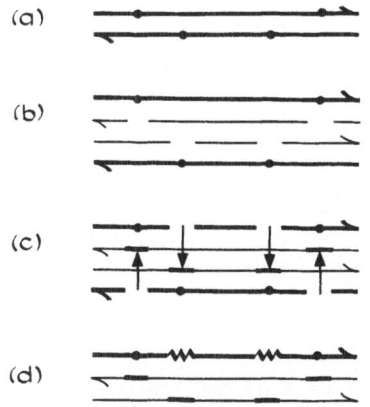

(a)

(b)

(c)

(d)

Fig.4. A general model for the postreplication repair of DNA damaged by UV radiation.
(a) Dots indicate photochemical lesions produced in the two strands of DNA. (b) DNA synthesis proceeds past the lesions in the parental strands leaving gaps in the daughter
strands. (c) Filling of the gaps in the daughter strands with material from the parental
strands by a recombinational process (depends
upon functional recA$^+$ genes). (d) Repair of
the gaps in the parental strands by repair replication. The reader is cautioned that steps
(c) and (d) are highly schematized, and will
probably be modified as additional data become available.

daughter strands. These gaps are then filled with material from the parental
strands by a recombinational process. The gaps formed in the parental strands
can be filled by repair replication.

This model is an oversimplification since it implies that the daughter-
strands are completely repaired during the first round of DNA replication.
However, it has been observed (20) that when parental-strand DNA is incorporated into the daughter-strands, it frequently carries radiation-produced
damage with it. This redistribution of the radiation damage between strands
implies that it may take several rounds of replication and of postreplication
repair before the DNA lesions are "diluted out", and a "viable" strand of DNA
is obtained. This idea is consistent with the observation that incubation
times equivalent to several generation cycles are required to eliminate the
effect of cyclobutane-type pyrimidine dimers on viability in an excision deficient strain of E. coli after a UV radiation fluence that leaves a surviving fraction of 0.65 (21).

On the basis of genetic and biochemical studies, postreplication repair
in bacteria has been divided into several independent pathways. The filling
of daughter-strand gaps is blocked in recA mutants (deficient in genetic recombination), but is only partially blocked in recB, uvrD, lexA and recF
strains. Chloramphenicol (CAP), an inhibitor of protein synthesis, also
partially blocks the filling of daughter-strand gaps, but not if the cells
are mutant at either lexA, recB, or uvrD, suggesting that in addition to
their independent functions, there must be another pathway that requires
the cooperation of all three gene products, and also requires the synthesis
of proteins after UV irradiation. This CAP-inhibitable pathway may be the
inducible error-prone pathway of postreplication repair that appears to be
important in UV radiation mutagenesis in excision deficient cells (14).

Repair of DNA Strand Breaks

The first step in the excision repair of UV radiation-produced DNA base
damage is the enzymatic production of single-strand breaks in DNA (Fig.3).
In addition to producing base damage, ionizing radiation also produces DNA
strand breaks by direct radiation chemical mechanisms. One might predict,

therefore, that there may be a considerable overlap in the enzymes required to repair these single-strand breaks even though they are produced by different mechanisms. This has been shown to be the case; the same genetic and physiological factors that control the repair of enzymatically produced DNA single-strand breaks after UV irradiation also control the repair of X-ray-produced DNA single-strand breaks (14).

DNA double-strand breaks appear to be unrepairable, and are lethal lesions in bacterial cells having fewer than two complete chromosomes per cell. In bacterial cells having two genomes, double-strand breaks can apparently be repaired by a process that requires functional recA$^+$ genes (22).

Inducible Error-Prone Repair

No mutants are detected in UV irradiated recA or lexA strains of E. coli. Since the recA and lexA genes control certain pathways of DNA repair, these results led to the conclusion that the molecular mechanism of mutagenesis is the error-prone repair of damaged DNA.

Since the recA and lexA genes also control certain functions that appear to be inducible by UV irradiation (e.g., filament formation, phage λ induction), this led to the suggestion that mutagenic repair (so-called "SOS" repair) that also depends upon recA and lexA genes is inducible (reviewed in 15,23).

The current model for inducible error-prone repair is the modification of one or more of the normal DNA polymerases by an inducible protein that inhibits the editing function of the DNA polymerases. In such a case, a DNA polymerase would incorporate random nucleotides opposite the damage, and thus produce errors (mutations) with a high probability. That an error-prone repair system is inducible seems clear, but few data are available to support this proposed molecular mechanism of "SOS" repair.

The error-free mechanisms of repair are considered to be largely constitutive, and would appear to include the major pathway of excision repair (since it can occur in buffer), and most of the pathways of postreplication repair (since they can proceed in the presence of CAP). Conversely , those pathways that are inhibitable by treatment with CAP (certain pathways in both excision and postreplication repair), imply the necessity of postirradiation protein synthesis, and thus are consistent with the process of induction.

There seems to be little doubt that a mistake made during the repair of damaged DNA is a molecular mechanism for mutagenesis. There is also a growing data base to suggest that mutagenesis is a molecular mechanism for carcinogenesis (reviewed in 24-26).

REFERENCES

1. Wang, S.Y. (ed.), 1976, "Photochemistry and Photobiology of Nucleic Acids", Academic Press, N.Y.
2. Huttermann, J., Kohnlein, W., Teoule, R., and Bertinchamps, A.J. (eds.), 1978, "Effects of Ionizing Radiation on DNA (Physical, Chemical and Biological Aspects)", Springer-Verlag, Berlin.
3. Hariharan, P.V. and Cerutti, P.A., 1977, Formation of products of the 5,6-dihydroxydihydrothymine type by ultraviolet light in HeLa cells, Biochemistry 16:2791-2795.

4. Wang, T.-C. V. and Smith, K.C., 1978, Enzymatic photoreactivation of Escherichia coli after ionizing irradiation: chemical evidence for the production of pyrimidine dimers, Radiat. Res. 76:540-548.
5. Rahn, R.O., 1979, Nondimer damage in deoxyribonucleic acid caused by ultraviolet radiation, Photochem. Photobiol. Rev. 4, 267-330
6. Smith, K.C. (ed.), 1976, "Aging, Carcinogenesis, and Radiation Biology (The Role of Nucleic Acid Addition Reactions)", Plenum Press, N.Y.
7. Rahn, R.O., Landry, L.C., and Carrier, W.L., 1974, Formation of chain breaks and thymine dimers in DNA upon photosensitization at 313 nm with acetophenone, acetone, or benzophenone, Photochem. Photobiol. 19:75-78.
8. Tyrrell, R.M., Ley, R.D., and Webb, R.B., 1974, Induction of single-strand breaks (alkali-labile bonds) in bacterial and phage DNA by near UV (365 nm) radiation, Photochem. Photobiol. 20:395-398.
9. Bonura, T., Smith, K.C., and Kaplan, H.S., 1975, Enzymatic induction of DNA double-strand breaks in γ-irradiated Escherichia coli K-12, Proc. Nat. Acad. Sci. USA 72:4265-4269.
10. Bonura, T. and Smith, K.C., 1975, Quantitative evidence for enzymatic-ally-induced DNA double-strand breaks as lethal lesions in UV irradiated pol$^+$ and polA1 strains of E. coli K-12, Photochem. Photobiol. 22:243-248.
11. Friedberg, E.C., Cook, K.H., Duncan, J., and Mortelmans, K., 1977, DNA repair enzymes in mammalian cells, Photochem. Photobiol. Rev. 2:263-322.
12. Hart, R.W., Hall, K.Y., and Daniel, F.B., 1978, DNA repair and mutagenesis in mammalian cells, Photochem. Photobiol. 28:131-155.
13. Roberts, J.J., 1978, The repair of DNA modified by cytotoxic, mutagenic, and carcinogenic chemicals, Adv. Radiat. Biol. 7:211-436.
14. Smith, K.C., 1978, Multiple pathways of DNA repair in bacteria and their roles in mutagenesis, Photochem. Photobiol. 28:121-129.
15. Hanawalt, P.C., Cooper, P.K., Ganesan, A.K., and Smith, C.A., 1979, DNA repair in bacteria and mammalian cells, Annu. Rev. Biochem. 48, 783 - 836 .
16. Hanawalt, P.C. and Setlow, R.B. (eds.), 1975, "Molecular Mechanisms for repair of DNA", Plenum Press, N.Y.
17. Hanawalt, P.C., Friedberg, E.C., and Fox, C.F. (eds.), 1978, "DNA Repair Mechanisms", Academic Press, N.Y.
18. Deutsch, W.A. and Linn, S., 1979, An apurinic DNA binding activity from cultured human fibroblasts that specifically inserts purines into de-purinated DNA, Proc. Natl. Acad. Sci. USA 76:141-144.
19. Lindahl, T., 1979, DNA glycosylases, endonucleases for apurinic/apyrimi-dinic sites and base excision repair, Progr. Nucleic Acid Res. Mol. Biol. 22, 135-192.
20. Ganesan, A.K., 1974, Persistence of pyrimidine dimers during postreplica-tion repair in ultraviolet light-irradiated Escherichia coli K-12, J. Mol. Biol. 87:103-119.
21. Ganesan, A.K., and Smith, K.C., 1971, The duration of recovery and DNA repair in excision deficient derivatives of Escherichia coli K-12 after ultraviolet irradiation, Molec. Gen. Genetics 113:285-296.
22. Krasin, F. and Hutchinson, F., 1977, Repair of DNA double-strand breaks in Escherichia coli, which requires recA function and the presence of a duplicate genome, J. Mol. Biol. 116:81-98.
23. Witkin, E.M., 1976, Ultraviolet mutagenesis and inducible DNA repair in Escherichia coli, Bacteriol. Rev. 40:869-907.
24. Kondo, S., 1976, Misrepair model for mutagenesis and carcinogenesis, in "Fundamentals in Cancer Prevention," (P.N. Magee et al., eds.) pp. 412-429, Univ. of Tokyo Press, Tokyo.
25. Trosko, J.E., and Chang, C.-C., 1978, Relationship between mutagenesis and carcinogenesis, Photochem. Photobiol. 28:157-168.
26. Trosko, J.E., and Chang, C.-C., 1978, The role of mutagenesis in carcino-genesis, Photochem. Photobiol. Rev. 3:135-162.

Light Induced Covalent Combination Between Furocoumarins and DNA. A Chemical Approach

G. Rodighiero

Istituto di Chimica Farmaceutica dell'Università
Padova, Italy

Furocoumarins are a group of organic compounds, in part naturally occurring in many plants especially of the families Um-
brelliferae and Rutaceae, in part prepared by chemical synthesis.

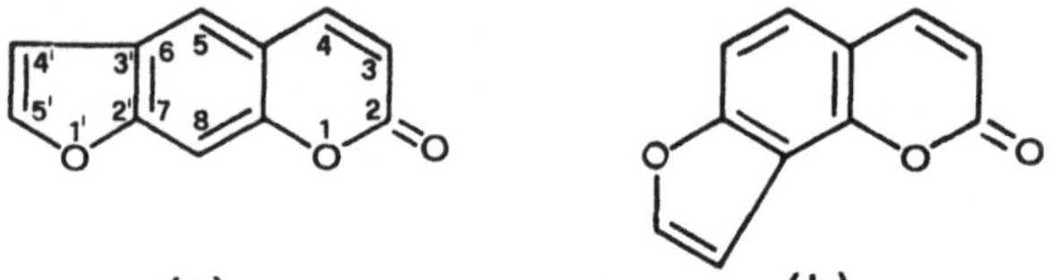

(a)   **(b)**

Fig:1 Molecular structures of psoralen (a) and angelicin (b).

From a chemical point of view, furocoumarins derive from the
condensation of a furanic ring with the coumarinic nucleus.This
condensation may occur in many different ways; therefore, various
isomer furocoumarins are possible. Among these, however, only
those represented in Fig.1 have practical importance. In the
first one, the condensation leads to a linear molecular structure;
in this case, the non-substituted furocoumarin is a naturally occurring compound called psoralen. Many other compounds derive
from it, either naturally occurring or prepared by synthesis;
they are called in general psoralens. The condensation of the
furanic ring with the coumarinic nucleus may lead also to an angular structure; in this case, the most simple compound is called angelicin or isopsoralen, and angelicins or isopsoralens
are called in general all furocoumarins deriving from it.
Furocoumarins, together with long wavelength ultraviolet radiation, are able to produce some biological effects. The effect
known since the longest time is the photosensitization of human
skin, leading to the production of erythema of various degree,
sometime accompanied by blisters; it is followed, after some
days, by a persistent dark pigmentation of the skin. In producing
this effect furocoumarins are very potent, being sufficient only
few µg of substance per cm^2 of skin and very small amounts of
radiation [1,2].

Other photosensitizing effects are killing of bacteria, inactivation of DNA viruses, inhibition of nucleic acids and protein synthesis in various kinds of cells, inhibition of the tumor transmitting capacity of Ehrlich ascites tumor cells [3,4].

Due to these properties, furocoumarins became interesting compounds from a biological point of view and especially for their therapeutic applications. Since long time furocoumarins have been used for obtaining the repigmentation of leukodermic spots of the skin, which are characteristic of vitiligo [5]. More recently, in some countries they have been used for the therapeutic treatment of psoriasis, mycosis fungoides and other skin diseases characterized by an increased reproduction of the cutaneous cells [6, 7]. The most videly used furocoumarin derivatives have been till now 8-methoxy-psoralen (xanthotoxin, 8-MOP, methoxsalen, meladinin), 5-methoxy-psoralen (bergapten, 5-MOP) and 4,5',8-trimethyl-psoralen (TMP, trisoralen).

As previously mentioned, furocoumarins display their biological and therapeutic effects only when associated with radiation. As they absorb only in the u.v. range, only ultraviolet radiation is effective; visible light doesn't produce any effect. The most effective radiations are those lying in the near ultraviolet, with wavelength in the range 320-400 nm; for a practical reason, the most used radiation has wavelength of 365 nm.

For their properties furocoumarins must be defined as potent photosensitizing compounds. Their mechanism of action has been widely investigated.

It is known that the common mechanism of action of many photodynamic compounds, such as hematoporphyrin, methylene blue, rose bengal, acridines, tiopyronin, is to produce an oxidation of the proteins contained in biological substrates. One of the demonstrated mechanisms of action consists in the absorption of a quantum of radiation and in the transfer of the absorbed energy to molecular oxygen, producing in such a way the very reactive singlet oxygen, which oxidizes the protein substrates. The production of singlet oxygen has been demonstrated by POPPE and GROSSWEINER [8] a few years ago also during irradiation of 8-methoxy-psoralen with long wavolength ultraviolet light. Singlet oxygen, however, is produced with a very low yield, much lower in comparison with other photodynamic compounds. Therefore, this mechanism of action can be cooperative, but, alone, it cannot explain the potent photosensitizing action of furocoumarins.

The most important way for the production of biological effects of furocoumarins is now attributed to a photobinding reaction that they can display with the pyrimidine bases of nucleic acids. This photoreaction has been discovered by MUSAJO et al. in 1964 and successively it was studied in many details [3,4]. I will summarize what now is known about this photoreaction, especially from a chemical point of view.

The first step of the interaction between furocoumarins and nucleic acids, in particular with native, double stranded DNA, is the formation of a molecular complex, due to the intercalation of a furocoumarin molecule between two base pairs of DNA. This is a dark-interaction, occurring by simple contact of the substances, without any irradiation. No covalent bonds are formed; only weak forces, mainly hydrophobic forces, are involved in the formation of this molecular complex [9].

The second step occurs by irradiation with long wavelength ultraviolet light; in this case, a chemical photoreaction takes place with formation of stable covalent bonds between furocoumarins and the pyrimidine bases of DNA.

This photoreaction has been clarified by working in a first time with the simple compounds. When a mixture of psoralen and one of the pyrimidine bases present in nucleic acids (thymine, cytosine or uracil) is irradiated with long wavelength ultraviolet light, a C_4-cyclo-addition between the two compounds takes place. In this photoaddition the 5,6-double bond of the pyrimidine bases is constantly involved. Psoralen and the other furocoumarins have two photo-reactive positions, namely the 3,4-double bound of the α-pyronic ring and the 4',5'-double bond of the furanic ring. In the photo-addition they can involve either one or the other position; therefore, two types of photoadducts can be originated: 3,4-photoadduct and 4',5'-photoadduct (Fig.2). For each photoadduct two structures are possible, deriving from the double possibility of addition of the molecules.

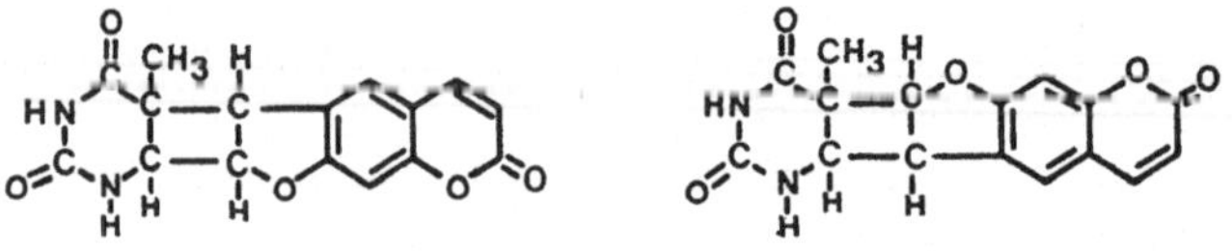

4',5'-photo-adducts

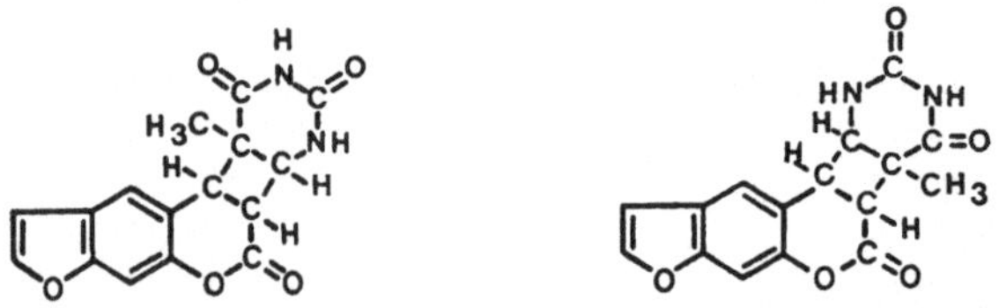

3,4-photo-adducts

Fig.2 Molecular structures of photoadducts between psoralen and thymine

Both 3,4- and 4',5'-photoadducts between psoralen and thymine have been isolated and characterized by MUSAJO et al.[10,11] after irradiation of aqueous solutions of the two compounds. The same photoadducts have been isolated also after irradiation of double stranded DNA in aqueous solution in the presence of psoralen, followed by hydrolysis of the same DNA by heating in aci-

dic medium [12]. Moreover, it was demonstrated that the same
photoaddition of furocoumarins to DNA takes place not only in
vitro , but also in vivo, that is inside the living cells [13,14].

In the psoralen molecule the 3,4-double bond is the most photoreactive position; in fact, when an aqueous solution of psoralen and thymine is irradiated at room temperature, that is in a
liquid condition, only 3,4-cyclo-addition takes place. The 4',5'-
cyclo-addition takes place only in some particular conditions,
for instance by irradiation of an aqueous solution of psoralen
and thymine in frozen condition, that is when the molecules are
immobilized in an ice matrix, or by irradiation of psoralen when
it is intercalated in DNA; generally, however, 4',5'-cyclo-adducts
are formed in a lower yield than 3,4-cyclo-adducts [15,16].

It has been demostrated, especially by SONG et al. [17], that
the excitation energy of the psoralen molecule, deriving from the
absorption of an ultraviolet photon, while in the singlet excited
state is distributed in the various positions of the molecule,
in the triplet excited state is very localized in the 3,4-posi-
tion, making therefore this position the most photoreactive. Therefore, we can assume that 3,4-cyclo-addition occurs prevalently
via triplet excited state.

Concerning the 4',5'-cyclo-addition, no clear evidence has been
till now obtained; however, on the basis of various considerations,
it was suggested that it occurs via singlet excited state.

The properties of the excited states of furocoumarins have been
recently studied by LAND et al. , using laser flash photolysis
[18, 19].

Till now only mono-adducts have been considered, that is compounds deriving from a cyclo-addition between one furocoumarin
molecule and one pyrimidine base. However, as furocoumarins have
two reactive sites, there is the possibility of a double photoaddition with the formation of a di-adduct between one furocoumarin molecule and two pyrimidine bases. Till now this possibility has never been verified by irradiation of the simple compounds
in aqueous solution; it occurs only by irradiation of psoralen
when it is complexed with native DNA.

Figure 3, representing the psoralen molecule intercalated between two base pairs in double stranded DNA, shows that a disposition of the psoralen molecule is possible in which the two reactive double bonds are both aligned with the 5,6-double bonds of the
two thymines, being in a suitable position to give the cyclo-
addition; therefore, a di-adduct can be formed without a great
distortion of the DNA structure [20].

The formation of the di-adduct is, of course, a two steps reaction. A first photon provokes the formation of a mono-adduct;
successively, the 4',5'-mono-adducts are able to absorb a second
photon and to give a second cyclo-addition with their 3,4-double
bond, forming thus a di-adduct.

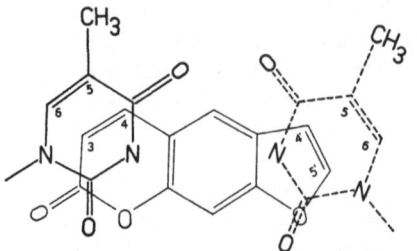

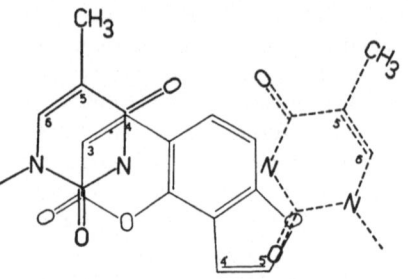

Fig.3 Projection of the psoralen molecule intercalated between two base pairs in DNA. Only two thymines are shown appertaining to the opposite strands

Fig.4 Projection of the angelicin molecule intercalated between two base pairs in DNA.

When a di-adduct is formed in native DNA, necessarily a cross-linkage between the two polynucleotidic strands takes place; in fact, the two reacting pyrimidines appertain to the opposite strands.

It must be pointed out that psoralen derivatives, that is furocoumarin derivatives having a linear molecular structure, generally form both mono-adducts and cross-linkages; by contrast, the isopsoralen derivatives are able to form only mono-adducts. In fact, due to their angular structure, it is impossible to find a position of the molecule intercalated between two base pairs of DNA in which its two reactive double bonds are both aligned with the 5,6-double bonds of two pyrimidine bases (Fig. 4). Therefore, angelicin and all its derivatives form only mono-functional adducts and no cross-linkages [20].

The photoreaction between psoralen and DNA can be summarized by the following scheme:

Free furocoumarin + nucleic acid ⇌ molecular complex

↓ hv 365 nm (Free furocoumarin)

↓ hv 365 nm (molecular complex)

photodimers, photooxidation products

photoaddition of the furocoumarin to a pyrimidine base

↓ hv 365 nm

further photoreaction of the furocoumarin already linked by its 4',5'-positions.

Although it is rather complicated, a complete kinetic analysis has been recently carried out by DALL'ACQUA et al. [16], using a mathematical model of the photoreaction. The most important results of this analysis showed that the main photoproduct is the 3,4-mono-adduct, thus confirming that the 3,4-double bond

of psoralen is the most photoreactive site of the molecule.
4',5'-mono-adducts are formed in a lower amount; they are partially transformed into cross-linkages. Di-adducts, or cross-
linkages, are constantly formed in a much lower yield, in respect to the mono-adducts.

I want now to stress the different significance, from a biological point of view, of the formation of mono-adducts and of
di-adducts.

Mono-adducts affect only one of the two polynucleotidic strands
of native DNA; therefore, the two strands of DNA can normally
separate, by denaturation or during the replication or transcription processes inside the cells; moreover, in this case DNA can
be repaired with a fast and simple mechanism which doesn't produce
errors in the sequence of nucleotides [21].

By contrast, di-adducts affect both polynucleotidic strands
of DNA forming inter-strand cross-linkages. Therefore, the two
strands of DNA, being covalently linked together, cannot be completely separated. Moreover, in their presence, DNA can also be
repaired, but with a much more complicated mechanism which may
be error prone, that is may produce some errors in the sequence
of nucleotides [21,22].

Therefore, we may understand the different significance, from
a biological point of view, of the formation of these two kinds
of damage of DNA. Generally, the formation of cross-linkages has
much stronger biological consequences; for instance, the erythema
produced on skin, is better correlatable with the formation of
cross-linkages, rather than with the formation of monoadducts.
Cross-linkages have generally a much stronger lethal effect on
cells than mono-functional adducts. AVERBECK et al.[23] have found
in yeast cells that cross-linkages produce rather nuclear mutations, while mono-adducts produce rather cytoplasmic "petite"
mutations.

Therefore, for biological investigations it is very important
to be able to produce, using furocoumarins, either only mono-
adducts or cross-linkages in DNA of living cells. This goal may
be achieved using different furocoumarin derivatives, for instance, an angelicin derivative, which forms only mono-functional
adducts, or a psoralen derivative which forms both mono-functional adducts and cross-linkages.

However, this goal has been recently achieved also using the
same psoralen derivative and performing the irradiation with very
short flashes of radiation, by means of a u.v.-laser. In fact,
as previously pointed out, the formation of a cross-linkage is
a two steps reaction; the first photon produces a mono-adduct,
which, by absorption of a second photon, gives rise to a further
photo-addition forming a cross-linkage. HEARST et al. [24] using
a water soluble psoralen derivative, demonstrated that irradiation
with a 15 nanoseconds pulse of a u.v.-laser produces in DNA only

mono-functional adducts; cross-linkages can be produced in DNA
by a successive irradiation with a second pulse of u.v.-laser.
    Therefore, the use of u.v.-laser, other than for studying the
properties of the excited states of furocoumarins, appears to be
a very useful means for addressing the photoreaction between
psoralen and DNA to the formation of only mono-adducts or cross-
linkages.
    Some biological experiments have been already worked out using
this procedure [25]; much further research work will be surely
performed in this field using u.v.-lasers.

References

1 - L.Musajo and G.Rodighiero - The skin-photosensitizing furo-
    coumarins - Experientia, 18, 153 (1962).
2 - M.A.Pathak, D.M.Krämer and T.B. Fitzpatrick - Photobiology
    and Photochemistry of furocoumarins (psoralens) - in
    Sunlight and Man, M.A.Pathak, L.C.Harber, M.Seiji and A.
    Kukita Eds., University of Tokyo Press, Tokyo, pg.335 (1976).
3 - L.Musajo and G.Rodighiero - Mode of photosensitizing action
    of furocoumarins - in Photophysiology, vol.VII, A.C.Giese
    Ed., Academic Press, New York and London, pg.115 (1972).
4 - L.Musajo, G.Rodighiero, G.Caporale, F.Dall'Acqua, S.Marciani,
    F.Bordin, F.Baccichetti and R.Bevilacqua - Photoreactions
    between skin-photosensitizing furocoumarins and nucleic
    acids - in Sunlight and Man, M.A.Pathak, L.C.Harber, M.Seiji
    and A.Kukita Eds. University of Tokyo Press, Tokyo, pg. 369
    (1976).
5 - T.B.Fitzpatrick,J.A.Parrish and M.A.Pathak - Phototherapy of
    vitiligo (idiopathic leukoderma)- in Sunlight and Man, M.A.
    Pathak, L.C.Harber, M.Seiji and A.Kukita Eds., University of
    Tokyo Press, Tokyo, pg. 783 (1976).
6 - J.A. Parrish, T.B.Fitzpatrick, L.Tanenbaum and M.A.Pathak,
    Photochemotherapy of psoriasis with oral metoxalen and longwa-
    ve ultraviolet light- New England J.Med. 291, 1207 (1974).
7 - K.Wolff, H.Hönigsmann, F.Gschnait and K.Konrad, Photochemo-
    therapie bei psoriasis - Deutsche Med. Wochenschrift, 100,
    2471 (1975).
8 - W.Poppe and L.I.Grossweiner, Photodynamic sensitization by
    8-methoxypsoralen via the singlet oxygen mechanism - Photo-
    chem. Photobiol. 22, 217 (1975).
9 - F.Dall'Acqua, M.Terbojevic, S.Marciani, D.Vedaldi and M.
    Recher, Investigation on the dark-interaction between furo-
    coumarins and DNA - Chem.Biol.Interactions, 21, 103 (1978).
10- L.Musajo, F.Bordin, G.Caporale, S.Marciani and G.Rigatti,
    Photoreactions at 3655 A between pyrimidine bases and skin-
    photosensitizing furocoumarins - Photochem. Photobiol. 6,
    711 (1967).
11- L.Musajo, F.Bordin and R.Bevilacqua, Photoreactions at 3655

A linking the 3,4-double bond of furocoumarins with pyrimidine bases - Photochem. Photobiol. 6, 927 (1967).

12- L.Musajo, G.Rodighiero, F.Dall'Acqua, F.Bordin, S.Marciani and R.Bevilacqua, Prodotti di fotocicloaddizione a basi pirimidiniche isolati da DNA idrolizzato dopo irradiazione a 3655 A in presenza di psoralene - Rend. Accad. Naz. Lincei (Rome), 42, 457 (1967).

13- L.Musajo, F.Bordin, F.Baccichetti and R.Bevilacqua, Psoralen-thymine C_4-cycloadducts formed in the photoinactivation with psoralen of Ehrlich ascites tumor cells - Rend. Accad. Naz. Lincei (Rome), 43, 442 (1967).

14- F.Dall'Acqua, S.Marciani, D.Vedaldi and G.Rodighiero, Formation of inter-strand cross-linkings in DNA of guinea pig skin after application of psoralen and irradiation at 365 nm - FEBS-Letters, 27, 192 (1972).

15- F.Dall'Acqua, S.Marciani, F.Bordin and R.Bevilacqua, Studies on the photoreaction (365 nm) between psoralen and thymine - La Ricerca Scientifica, 38, 1094 (1968).

16- F.Dall'Acqua, S.Marciani, F.Zambon and G.Rodighiero, Kinetic analysis of the photoreaction (365 nm) between psoralen and DNA - Photochem. Photobiol. 29, 489 (1979).

17- P.S.Song and K.J.Tapley, Photochemistry and photobiology of psoralens - Photochem. Photobiol. 29, 1177 (1979).

18- R.V.Bensasson, E.J.Land and C.Salet, Triplet excited state of furocoumarins: reaction with nucleic acid bases and amino acids - Photochem. Photobiol. 27, 273 (1978).

19- E.J.Land and T.G.Truscott, Triplet excited state of coumarin and 4',5'-dihydropsoralen: reaction with nucleic acid bases and amino acids - Photochem. Photobiol. 29, 861 (1979).

20- F.Dall'Acqua, S.Marciani, L.Ciavatta and G.Rodighiero, Formation of inter-strand cross-linkings in the photoreactions between furocoumarins and DNA - Z. Naturforsch. 26b, 561 (1971).

21- R.S.Cole and R.R.Sinden, Psoralen cross-links in DNA: biological consequences and cellular repair - in Radiation Research, O.F.Nygaard, H.I.Adler and W.R.Sinclair Eds., Academic Press, New York and London, pg. 582 (1975).

22- F.Bordin, F.Carlassare, F.Baccichetti and L.Anselmo, DNA repair and recovery in Escherichia coli after psoralen and angelicin photosensitization - Biochim. Biophys. Acta, 447, 249 (1976).

23- D.Averbeck, R.K.Biswas and P.Chandra, Photoinduced mutations by psoralens in yeast cells - in Photochemotherapie, E.G. Jung Ed., Schattauer Verlag, Stuttgart, pg. 97 (1975).

24- B.H.Johnston, M.A.Johnson, C.B.Moore and J.E.Hearst, Psoralen-DNA photoreaction: controlled production of mono- and diadducts with nanosecond ultraviolet laser pulses - Science, 197, 906 (1977).

25- S.P.Peterson and M.W.Berns, Effect of psoralen and near U.V. on vertebrate cells in culture: comparison of laser with standard lamp - Photochem. Photobiol. 27, 367 (1978).

Part II

Photodynamic Therapy of Tumors

The Molecular Biology of Photodynamic Action

G. Jori

Centro C.N.R. Emocianine, Istituto di Biologia Animale, Università di Padova
35100 Padova, Italy

1. Introduction

According to BLUM [1], the expression photodynamic action indicates those photosensitized reactions, induced in biological systems by visible light, in which molecular oxygen is consumed.
This operational definition is still valid, although a few photosensitizers, which are widely diffused in living organisms
e.g., flavins, ketones, psoralens), may not require oxygen for
their action. On this basis, photodynamic action is promoted
only by light wavelengths greater than about 320 nm; more energetic wavelengths would be directly absorbed by cell constituents, especially proteins and nucleic acids, which might then
undergo photochemical modifications even in the absence of photosensitizing compounds [2]. Thus, photodynamic action is induced by the primary interaction of the incident light with the
photosensitizer molecule. The sequence of events, following the
absorption of a photon by the sensitizer, is illustrated in the
following scheme:

$$
\begin{array}{lll}
^{o}S + h\nu & \longrightarrow & ^{1}S & \text{light absorption} \\
^{1}S & \longrightarrow & ^{o}S & \text{non-radiative singlet decay} \\
^{1}S & \longrightarrow & ^{o}S + h\nu^{\bullet} & \text{fluorescence emission} \\
^{1}S + ^{o}S & \longrightarrow & 2\ ^{o}S & \text{self-quenching of } ^{1}S \\
^{1}S & \longrightarrow & ^{3}S & \text{intersystem crossing} \\
^{3}S & \longrightarrow & ^{o}S & \text{non-radiative triplet decay} \\
^{3}S & \longrightarrow & ^{o}S + h\nu'' & \text{phosphorescence emission} \\
^{3}S + ^{o}S & \longrightarrow & 2\ ^{o}S & \text{self quenching of } ^{3}S
\end{array}
$$

where ^{o}S, ^{1}S and ^{3}S represent the ground state, the first excited
singlet state, and the first excited triplet state of the photosensitizer, respectively. The radiative decay of the triplet state (phosphorescence emission) is usually important only in rigid
matrices, whereas stages such as triplet-triplet annihilation are
unimportant at the light intensities supplied by conventional
irradiation sources.

Out of the electronically excited derivatives of the photosensitizer, the triplet state is generally endowed with greatest
photoreactivity [3]; this fact is a consequence of its relatively

long lifetime (1 ms to some seconds), whereas the lifetime of ^1S is too short (ca. 1 ns to 0.1 μs) to give it a good chance of interacting with other molecules before decaying back to the un-reactive ground state. Therefore, one requisite of efficient pho-todynamic sensitizers is the possess of a high quantum yield of intersystem crossing to the lowest triplet state.

2. Mechanisms of Photodynamic Processes

There are two main pathways which are open to the triplet sensi-tizer. In the so-called type I mechanism [4], the ^3S species re-acts directly with the substrate molecule D to give either hy-drogen or electron transfer. In the event that a reducing substra-te is present, this reaction pathway may be represented as fol-lows:

$$^3S + D \longrightarrow SH\cdot + D\cdot \tag{1}$$

or

$$^3S + D \longrightarrow S\cdot^- + D\cdot^+ \tag{2}$$

As a rarer alternative, the ^3S species can transfer one electron to oxygen, generating the superoxide anion $O_2\cdot^-$ [5]. In most cases, the ground state of the photosensitizer is reformed from the se-mireduced form by reaction with molecular oxygen.

On the other hand, the free radical-species derived from the substrate through stages (1) and (2) may give a wide variety of possible further reactions: e.g., promotion of chain processes by interaction with other substrate molecules or reaction with oxygen to yield a stable fully oxidized product. One example of type I photodynamic process is provided by the photooxidation of cysteine to cysteic acid, sensitized by crystal violet or by other triphenylmethane dyes [6]:

$$RSH + S^* \longrightarrow RS\cdot + SH\cdot$$
$$RS\cdot + O_2 \longrightarrow RSO_2\cdot$$
$$RSO_2\cdot + O_2 \longrightarrow RSO_4\cdot$$
$$RSO_4\cdot + O_2 \longrightarrow R\text{-}SO_4H + RS\cdot$$
$$R\text{-}SO_4H + H_2O \longrightarrow RSO_3H + H_2O_2$$

where RSH represents a cysteine molecule. The photoprocess in-volves the abstraction of the thiol hydrogen by photoexcited dye S^* with the formation of a thiyl radical, which in turn o-riginates a chain reaction.

In the type II mechanism [5], ^3S transfers the excitation e-nergy to molecular oxygen, which is a triplet in its ground sta-te and is promoted to the lowest excited singlet state:

$$^3S + {}^3O_2 \longrightarrow {}^\circ S + {}^1O_2 \tag{3}$$

The species of $^1O_2$ is highly electrophilic and can attack electron-rich sites of biomolecules; a well known example is given by the photooxidation of the amino acid methionine to methionine sulfoxide [7]:

$$R-S-CH_3 + {}^1O_2 \longrightarrow R-SO-CH_3$$

The overall photoreaction, although apparently simple, has been shown to involve several distinct intermediates [7].

In general, in a given photodynamic system, both types of mechanisms may competitively occur. As one deduces from equations (1) or (2) and (3), the rate of a typical type I photoprocess is given by:

$$v_I = k_I [{}^3S][D] \qquad \text{and} \qquad v_{II} = k_{II} [{}^3S][O_2] \qquad (4)$$

Thus, the efficiency of either mechanism is controlled by the nature of the sensitizer and the substrate, by the relative concentration of oxygen and substrate, and by the rate constants for the substrate-sensitizer and oxygen-sensitizer interaction. Since oxygen is much less soluble in water than in most organic solvents, decreasing the polarity of the solvent is expected to favour the type II photosensitization mechanism; moreover, the lifetime, hence the reactivity, of singlet oxygen increases up to two orders of magnitude going from water to aprotic solvents [8]. On the other hand, the complexation of the sensitizer with the substrate, prior to irradiation, which often takes place in biological systems, enhances the probability of a type I mechanism [9].

Therefore, the definition of the mechanism of a given photosensitized reaction is a necessary step in the interpretation of any photodynamic phenomenon. It is worthwhile emphasizing that a type I photoprocess usually leads to the modification of substrates in close proximity of the photoexcited sensitizer, while singlet oxygen may diffuse over appreciable distances before reacting or being deactivated.

A familiar method for distinguishing type I and type II mechanisms is the use of potential inhibitors or enhancers, which are known to act at specific stages of the photoprocess by interacting physically or chemically with the various intermediates. For example, the lifetime of singlet oxygen increases almost tenfold in D_2O with respect to H_2O [8]; hence, the kinetics of a photoreaction proceeding via singlet oxygen should be consistently increased using D_2O as a solvent [5, 10]. Moreover, azide is a very efficient quencher of singlet oxygen, and its addition to the irradiated system lowers the reaction rate if the latter species is the active intermediate [5]. Another clean method to detect singlet oxygen or free radical species is provided by the identification of the photooxidation end-products.

Table 1   Effect of the Photosensitization Mechanism on the
Photooxidation Products of Specific Substrates

Substrate	Type I Products	Type II Products
Cholesterol	7α-hydroperoxyde	5α-hydroperoxide
Cysteine	Cysteic acid	Cystine
Methionine	Methional	Methionine sulfoxide

As shown in Table 1, different photoproducts have been isolated
from some biomolecules depending on the photosensitization mechanism.
A detailed discussion of these subjects is given in references
[3] and [5].

3. The Chemistry of Photodynamic Sensitizers

A very large number of organic and inorganic compounds have be-
en reported to act as photodynamic sensitizers [11]. An important
distinction is between endogenous and exogenous photosensitizers,
i.e. between photosensitizing moieties which are naturally present in biological systems or which may be artificially added
to the irradiated system.

Table 2   Examples of Endogenous Photosensitizers

Photosensitizer	Chemical structure	Absorption ranges
Psoralens	Condensation of a co-umarin nucleus with a furanoic ring	320–380 nm
Hypericin	Octacyclic polyhydro-xylated hydrocarbon	360–420 nm
Porphyrins	Cyclic tetrapyrroles eventually coordina-ted with metal ions	Ca. 400 nm (very in-tense), minor abs. at longer wavelength
Bilirubin	Linear tetrapyrrole	420–480 nm
Flavins	N-containing trihete-rocyclic compounds	400–460 nm
Pyridoxal and its analogues	Poly-substituted py-ridines	320–400 nm, very pH-dependent
Metal ions	Cu^{++}/O_2system Fe^{+++}	330–400 nm broad above 500 nm

Some important endogenous photosensitizers are listed in Table 2. Apparently, these photosensitizers can absorb different portions of the visible spectrum, often causing harmful effects to living systems, owing to the critical biological functions performed by them.

As concerns exogenous photosensitizers, many of them are tri-heterocyclic compounds (see Table 3). However, no clear-cut relationship between chemical structure and photodynamic efficiency is evident at present. In general, on the basis of the previously outlined mechanistic features of these photoreactions, one might anticipate that those structural parameters, which enhance the quantum yield of triplet state production, should also enhance the photodynamic activity of photosensitizers. Thus, WADE and SPIKES [12] demonstrated that the introduction of heavy atoms (e.g., bromine or iodine) into the fluorescein nucleus increases the probability of singlet-triplet intersystem crossing with a parallel increase of the photodynamic efficiency. Analogously, Cauzzo et al. [13] ascertained that the presence of coordinated paramagnetic metal ions inhibits the photosensitizing activity of porphyrins owing to a drastic reduction of the triplet lifetime; diamagnetic ions, such as Mg^{++} which is present in chlorophyll, display a greatly reduced quenching effect.

Table 3 Examples of Exogenous Photosensitizers

Photosensitizer	Chemical structure	Absorption range
Acridine orange	Dimethylamino-acridine	460–510 nm
Eosin Y	Tetrabromoxanthene	490–530 nm
Methylene blue	Dimethylamino-phenothiazine	630–690 nm
Metal ions	Uranyl UO_2^+	400–500 nm
	Ruthenium bipiridyl	530–570 nm

The presence of photosensitizing compounds in living organisms is responsible of many phenomena of natural photosensitivity. Thus, photoactivation of psoralens often results in skin erythema [14], whereas alteration in the metabolism of porphyrins may cause different forms of the disease known as porphyria [15]. However, in most cases, living systems are efficiently protected against the photodynamic damages induced by endogenous photosensitizers. Thus, in hemoproteins, the porphyrin group is associated with a paramagnetic Fe^{+++} ion, which eliminates any photodynamic activity. On the other hand, in photosyn-

thetic organisms, the possibility of chlorophyll-induced photodynamic damage is minimized by the presence of high concentrations of carotenes, which are strong quenchers of both singlet
oxygen and triplet chlorophyll [5].

4. Photodynamic Processes "in vitro"

The molecular sites of the fundamental cell constituents, which
are most readily susceptible to photodynamic modification, are
shown in Table 4. As discussed previously, these targets are
endowed with appreciable electron density, which underlines the
electrophilic character of photodynamic processes. One serious
drawback still existing in this section of photodynamic studies
is the scarcity of information about the photoproducts of some
substrates. Moreover, in some cases, the initial photoreaction
products (e.g., peroxides) may undergo further (photo)modifica-
tions, eventually involving other biomolecules.

Table 4 Most Readily Photooxidizable Sites of Biomolecules

Biomolecule	Attacked site(s)	Main end-products
Lipids	Carbon-carbon double bonds	Peroxides
Steroids	Cholesterol	Hydroperoxide
Polysaccha-rides	Hydroxyl groups in a type I mechanism	"Onic" acid and ketones
Nucleic acids	Guanine moiety	Destruction of the purine ring
Proteins	Tryptophan	Kynurenine derivatives
	Histidine	Destruction of the imidazole ring
	Tyrosine	Unknown
	Methionine	Methionine sulfoxide Methional
	Cysteine	Cystine, cysteic acid

Notwithstanding the variety of potentially photolabile sites
and of intermediate and final photoproducts, the course of a
photodynamic process can often be adequately controlled. One
approach, which may be pursued to direct the photoreaction toward specific predetermined targets, involves the control of
the ionic state of the substrate: as a consequence of the electrophilic character of most photodynamic processes, protonation
of biomolecules having electron-rich functions results in a reduction of their photosensitivity [16]. A related approach takes advantage of the strong influence of the experimental condi-

tions on the activity of many photosensitizers: a thorough analysis of these parameters led Jori et al. [17] to find irradiation conditions allowing the selective photooxidation of the cysteinyl, methionyl, and tryptophyl residues in a protein.

A different approach utilizes the screening effect exerted by the three-dimensional conformation of a macromolecule to the interaction between the photooxidizing agents and the susceptible sites [18]; thus, in a protein, the amino acid side chains, which are close to the external surface of the molecule, will react faster with either singlet oxygen or the triplet dye than the residues which are buried in internal regions. The determination of the rate constant for the photomodification of the various residues yields a quantitative estimation of their degree of burial within the protein matrix.

A final approach is represented by the irradiation of covalent or non-covalent complexes between the sensitizer and the biomolecule [19]. In this way, only the susceptible moieties nearest to the bound photosensitizer will be affected, at least in the early stages of the photoreaction. This technique was applied to study the topography of specific region of protein molecules, especially hemoproteins and pyridoxal-dependent enzymes, as well as to map the architecture of complex biological structures [9, 16].

Even if the photosensitizer is only weakly bound with the substrate, localized photodynamic effects can be obtained, provided the absorption spectrum of the bound dye is shifted with respect to that of the free dye [20]. Thus, acridines bind to DNA either by stacking on the outside of the molecule (absorption maximum at 465 nm) or by intercalation between DNA bases (absorption maximum at 502 nm). Illumination of the complex at 470 nm [21] produced singlet oxygen with greater efficiency than 502 nm-irradiation; clearly stacked dye molecules are better located for interacting with oxygen. On the other hand, irradiation with light preferentially absorbed by the intercalated dye induced mutations with good yield owing to the high probability of direct interaction of the excited dye with the nucleic acid bases.

5. Photodynamic Processes "in vivo"

The basic concepts and mechanisms, which have been discussed for "in vitro" photodynamic processes, apply also "in vivo". Thus, chemical modification studies performed on organelles isolated from irradiated cells in the presence of photosensitizers indicate that the targets, which appear to be most photolabile from investigations with the isolated biomolecules, are readily photooxidized also "in vivo" [3, 22]. Moreover, in systems involving the presence of singlet oxygen at the cellular level, one observes again the protective action by azide and the enhancement of the photodynamic effects by D_2O [21].

In a recent review [23], ITO described the problems which are created by the different possible locations and motions of sensitizers and quenchers in cells. In general, three modes of photodynamic action can be distinguished, whose relative importance depends mainly on the physico-chemical properties of the dye used.

- 1°: Cationic dyes (e.g., thiazines, including methylene blue and thiopyronine) usually remain outside the cell. The singlet oxygen mechanism gives an important contribution to the overall photodynamic process, and the damage is mainly located at the membrane level as a consequence of attack on membrane proteins and unsaturated lipids. However, photoinduced cross-linking of membrane proteins has also been reported.
- 2°: Fluoresceins (e.g., rose bengal, eosin Y) and, possibly, porphyrins locate in the cytoplasm. Visible light-illumination causes the damage of cytoplasmic structures, owing to attack both on enzymes and on RNA in type I and/ or type II photoprocesses.
- 3°: Acridines and other dyes bind to DNA in the nucleus of the cell; therefore, in this case, photodynamic damage principally involves DNA with the consequent frequent induction of genetic alterations.

Apparently, the situation may be different for any given system, since small variations in the experimental conditions often influence the competition between type I and type II mechanisms. Even the assumption that singlet oxygen may diffuse and damage cellular components over relatively large distances from its generation site, while type I mechanisms require close contact between substrate and sensitizer may not be generally valid. Actually, recent studies in our laboratory showed that, in the case of hematoporphyrin photosensitization in aqueous dispersions of anionic micells, electron transfer from the photoexcited dye to the substrate can take place even if the two moieties are located on different sides of the charged layer of the micelle.

In any case, the above cited investigations open the way to the control of the sites attacked and to the predetermination of the type of damage induced in a cell by an appropriate choice of the photosensitizer and of the reaction conditions. Exciting prospects are the investigation of the structure and function of supramolecular systems by photodynamic techniques, as well as the improvement of the present methods of photoprotection and phototherapy. The recently developed approaches for the photoprotection of porphyric patients and for the phototherapy of neonatal jaundice, psoriasis and several kind of tumours are stimulating examples in this direction.

6. References

1. Blum F.H., 1964, "Photodynamic Action and Diseases Caused by Light", Hafner Publ., New York.

2. Smith K.C. and Hanawalt P.C., 1969, "Molecular Photobiology", Academic Press, New York.
3. Spikes J.D. and Livingston R., 1969, Advan. Radiat. Biol. 3, 29-121.
4. Gollnick K., 1968, Advan. Photochem. 6, 1-3.
5. Foote C.S., 1976, in "Free Radicals in Biological Systems" (W.A. Pryor, ed.) Vol. II, pp. 1 5-133, Academic Press, New York.
6. Gennari C., Cauzzo G. and Jori G., 1974, Photochem. Photobiol. 20, 497-500.
7. Sysak P.K., Foote C.S. and Ching T.Y., 1977, Photochem. Photobiol. 26, 19-27.
8. Merkel P.B., Nillson R. and Kearns D.R., 1972, J. Am. Chem. Soc. 94, 1030-1031.
9. Jori G., 1974, Anais Academia Brasileira de Ciencias 45, 33-44.
10. Nillson R., Merkel P.B. and Kearns D.R., 1972, Photochem. Photobiol.. 16, 117-124.
11. Santamaria L. and Prino G., 1964, in "Research Progress in Organic, Biological and Medicinal Chemistry" (U. Gallo and L. Santamaria, eds.) Vol. 1, pp. 260-336, Società Editoriale Farmaceutica, Milano.
12. Wade M.J. and Spikes J.D., 1971, Photochem. Photobiol. 14, 221-274.
13. Cauzzo G., Gennari G., Jori G. and Spikes J.D., 1977, Photochem. Photobiol. 25, 389-395.
14. Musajo L. and Redeghiero G., 1972, "Photophysiology" (A.C. Giese, ed.) Vol. 7, pp. 115-147, Academic Press, New York.
15. Harber L.C., Baer L., Bickers D.R., 1974, in "Sunlight and Man" (T.B. Fitzpatrick et al., eds.) pp. 631-653, University of Tokyo Press, Tokyo.
16. Spikes J.D. and MacKnight M.L., 1970, Ann. N.Y. Acad. Sci. 171, 149-162.
17. Jori G., Folin M. and Galiazzo G., 1976, La Chimica e l'Industria 58, 486-492.
18. Ray W.J. and Koshland D.E., 1962, J. Biol. Chem. 237, 2493-2505.
19. Jori G. and Spikes J.D., 1978, in "Photochem. Photobiol. Reviews" (K.C. Smith, ed.) Vol. 3, pp. 193-275, Plenum Press, New York.
20. Jori G., Galiazzo G. and Scoffone E.,1971, Experientia 27, 379-381.
21. Kobayashi K. and Ito T., 1977, Photochem. Photobiol. 25, 385-388.
22. Spikes J.D., 1977, in "The Science of Photobiology" (K.C. Smith, ed.) pp. 87-112, Plenum Press, New York.
23. Ito T., 1978, Photochem. Photobiol. 28, 493-508.

Photoradiation Therapy of Malignant Tumors; Role of the Laser

T.J. Dougherty, R.E. Thoma, D.G. Boyle, and K.R. Weishaupt

Division of Radiation Biology, Roswell Park Memorial Institute
Buffalo, NY 14263, USA

Introduction

Photoradiation therapy, in which hematoporphyrin derivative (Hpd) is activated by visible light in situ is being investigated for the control of solid malignant tumors in man |1,2|. This method is based upon the ability of Hpd to first accumulate and be retained in malignant tissue to a greater degree than in some normal tissues |3,4,5|, and then to produce singlet oxygen ($^1O_2$) when activated by visible light, thus producing a cytotoxic effect |6|. While this latter property, known as a photodynamic effect |7|, is common to a wide range of materials, its combination with tumor localizing ability is quite unique. Tetraphenylporphine sulfonate appears to possess similar properties to Hpd but suffers the disadvantage of prolonged retention in serum and tissues |8,9|, thus rendering animals photosensitive for extended periods. This is also a problem with Hpd |1,2| but is of relatively short duration (approximately 30 days, depending on dose) and can be avoided by preventing sunlight exposure. We have examined many other porphyrins as potential tumor photosensitizers in animals, but none has been found to be as effective as Hpd |10|.

To date, most of the patients treated by this technique have had cutaneous and/or subcutaneous lesions in which the Hpd was photoactivated by externally applied red light (600-700nm). While red light penetrates tissue to a greater extent than any other visible wavelengths |1|, its fall-off with depth is rapid with an effective penetration depth of approximately 2cm |1,2|. This light may be obtained from various conventional lamps or from an appropriate laser. Since relatively high intensity between 625 and 640nm is necessary |1,11|, a dye laser provides the only feasible source. While the He-Ne laser emits at an appropriate wavelength (632.8nm), the output of current equipment (50mw) is too low for most clinical applications.

The major advantage to the laser, however, is not intensity or wavelength, but the flexibility of use, especially when the output is delivered through a fiber optic. Currently available glass fibers allow coupling to the laser with high efficiency and very low attenuation of power with fiber length. Thus, the necessary activating light for photoradiation therapy can be delivered through various endoscopes or directly into tumor masses through inserted needles (interstitial implant). Early results, in animals and humans, using this latter technique are discussed in this report.

Materials and Methods

Hematoporphyrin derivative (Hpd): Hematoporphyrin hydrochloride (Roussel Corp., New York, N.Y.), 1 part, was dissolved in glacial acetic acid: sulfuric acid (19:1, by volume) and allowed to stand overnight at room temperature.

The mixture was filtered by gravity through a Whatman No. 1 filter paper (1 to 3 hours depending on total volume) and neutralized to approximately pH 6.0 by addition of 3% sodium acetate solution. The precipitated material was recovered by filtration and washed extensively with distilled water until the washings were neutral. The solid was then dried in a vacuum to remove all traces of acetic acid and water and was stored in the dark in the solid state at -20°. This material had 3 components detectable by thin-layer chromatography (Brinkman Polygram SIL-N-HR plates; benzene:methanol, 4:1, by volume) in a ratio of 15, 30 and 55% in increasing R_F. The material with the lowest R_F (15%) corresponds to a major peak in the crude hematoporphyrin hydrochloride. The other 2 components have been identified as a monoacetate (30%) and diacetate (55%) by using $|^{14}C|$acetic acid to prepare Hpd and by isolating the individual components by thin-layer chromatography.

An injectable solution of Hpd was prepared by mixing 1 part Hpd with 50 parts by volume of 0.1 N sodium hydroxide and stirring at room temperature for 1 hour. The pH was brought to 7.2 to 7.4 by addition of 0.1 N hydrochloric acid (approximately 15 parts), made isotonic by addition of sodium chloride, and finally brought to 200 parts total with 0.9% sodium chloride solution. The final solution of 5mg/ml was sterilized by Millipore filtration and tested for sterility and pyrogenicity. The solution was stored in the dark at -20° until used.

Dye Laser: A Spectra-Physics Model 375 dye laser, pumped either by a 4 or 15-watt Spectra-Physics argon laser was used in all cases. Rhodamine B was used in the dye laser which was operated without the tuning wedge. This system emits energy between 625 and 640nm and allows for increased output compared to that of a single line obtained with the wedge in place. The output from the dye laser was directed into the lens of the fiber-optic coupling system which consisted of a spatial filter (Oriel Corp., Stamford, Connecticut) in which the pinhead hole was replaced by the shaft of a 20-gauge needle permanently mounted in the pinhole holder. The entire assembly was mounted on a X,Y,Z translator which, in turn, was mounted on a rotating platform. A Quartz Silice (Paris, France), 200μm core, glass clad, step-index fiber (QSF-A-200, Quartz Products, Plainfield, New Jersey) was fitted into the coupling system after both ends had been cleaved to an optically flat surface using a sapphire knife wedge (Math Associates, Great Neck, New York). Coupling of this fiber to the laser was near 80%. Overall efficiency (dye laser output through the fiber, compared to argon laser output) was approximately 10%. Figure 1 is a schematic of the system.

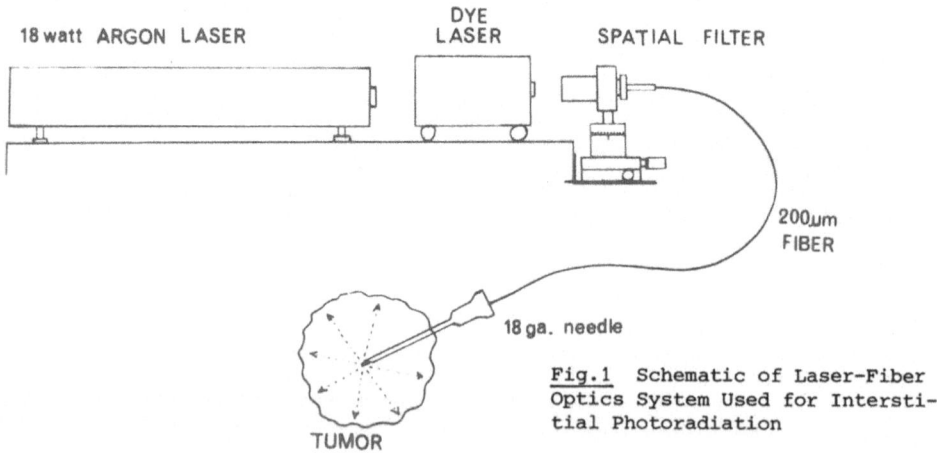

Fig.1 Schematic of Laser-Fiber Optics System Used for Interstitial Photoradiation

Procedures

Primary tumors in animals: Spontaneous tumors in pet cats and dogs were obtained directly from owners and were generally returned following treatment. Among the eight animals, were three with osteogenic sarcoma, one with fibrosarcoma, one with malignant melanoma, one adenocarcinoma, one squamous cell carcinoma and one with mastocytoma. All were biopsy confirmed. The animals were injected intravenously with 5.0mg/kg Hpd and treated with light, while under general anesthesia, two or three days later. One of the animals received externally applied light only, and one received both external and internally applied light from the laser. In these cases, the laser beam, delivered through the fiber, was directed to the tumor with a margin of 5mm. Intensity ranged from 30 to 120mw/cm^2 for 10 to 20 minutes.

For fiber insertion, the tip was soaked 5-10 minutes in antiseptic solution (Cidex), washed with alcohol and threaded through an 18-gauge needle which had been inserted directly into the tumor to the desired position. It was important to avoid bleeding through the needle since this tended to cause clotting on the tip of the fiber with consequent loss of light. The light was delivered at intensities of 180-400mw (total output) for 20 to 40 minutes. In general, if tumors exceeded 1cm in maximum diameter, multiple insertions were used with the needles placed approximately 1cm apart. The treatments were carried out consecutively.

Tumors in humans: All patients had histologically confirmed malignancy which had recurred or not responded to various conventional treatments. Of the five patients treated by fiber insertion (interstitial implant), three had soft tissue recurrence of breast carcinoma, one a recurrent mixed salivary gland carcinoma in the head and neck region and one a recurrence of rectal carcinoma involving the labia. All patients understood the experimental nature of the treatment and signed consents. The patients received Hpd intravenously in doses of 2.5 to 5.0mg/kg. At this time and several times thereafter, they were strongly cautioned to remain out of direct sunlight indoors and outdoors. From three to twelve days following injection, tumors were treated by the laser by inserting the fiber through an 18-gauge needle which had been positioned in the desired area. In some cases, a local anesthetic (Lidocaine) was injected before the fiber was placed. Total intensities from the fiber ranged from 150 to 400mw and treatment times ranged from 20 to 60 minutes. Since in all cases the tumors exceeded 1cm in maximum diameter, multiple consecutive insertions were used with the fiber placed 1-4cm apart.

Results

Table 1 is a summary of current results in animal tumors treated by 625-640nm light from the dye laser. All but one (Dog #7) were treated by an interstitial implant of the light delivery fiber directly into the tumor mass. When tumors exceeded 1cm diameter, multiple implants were used to cover the entire tumor volume. In general, the fibers were placed 1-2cm apart. All tumors, with the exception of a large osteogenic sarcoma involving the tibia of Dog #5, responded to treatment. That this lack of response is probably more related to size than resistance of this type of tumor is indicated by total erradication of an osteogenic sarcoma in the mandible of a cat (#1) which has now gone more than two years without local recurrence or metastatic disease. A similar type of tumor, in a dog, involving the maxillary sinus and hard palate has also undergone apparent total regression (radiologically), although the follow-up period is still short (2 months). In both cases, bone remodeling has occurred in the area of previous tumor.

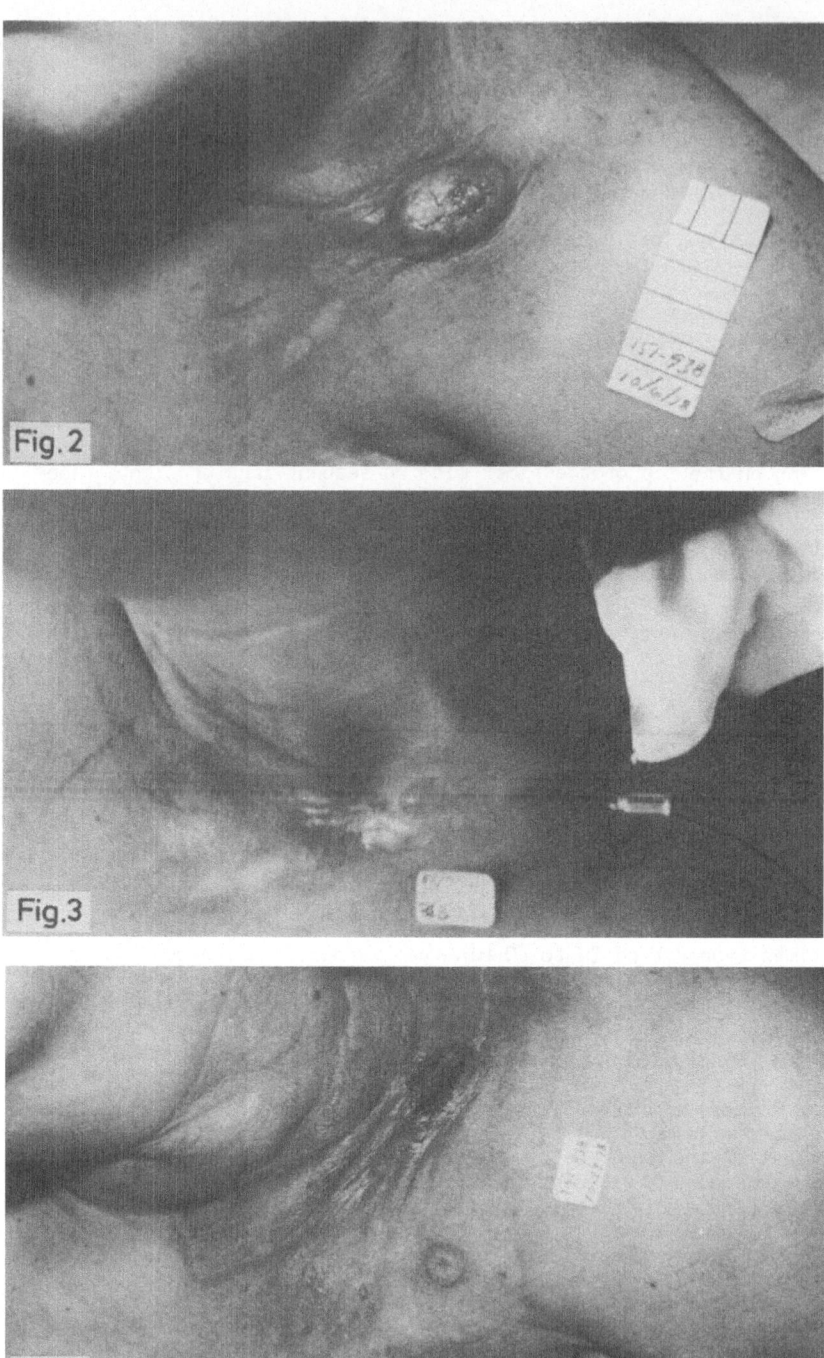

Figs. 2-4. Figure captions see opposite page

Control of a highly pigmented malignant melanoma lesion in Dog #3 proved to be more difficult than for a non-pigmented lesion in the same animal. While the latter was erradicated after two treatments, the former continued to recur. It was finally erradicated by surgery, followed several weeks later by externally applied photoradiation to the remaining mass.

In the remaining animals, follow-up was incomplete and is still too short for meaningful conclusions except to say that response is evident.

The results of the five patients treated to date, by the interstitial implant technique are summarized in Table 2. In several cases, follow-up time is too short for definitive conclusions. The most responsive tumors were metastatic adenocarcinoma of the breast. Figure 2 indicates such a tumor measuring 4cm x 2cm diameter in patient #3. Four days following injection of 3.5mg/kg Hpd, the fiber was inserted into both the anterior and posterior sections of the tumor, consecutively, and 300mw total emitted power from the fiber was applied for 30 minutes in each position (Figure 3). This procedure was repeated two days later. The mass gradually reduced over a period of several days to the point where the nodule appeared completely flat (Figure 4). Several 2-3cm nodules on the back of this patient (more or less contiguous over an area of approximately 10cm x 13cm) were treated similarly except each individual nodule received a single treatment. These lesions also appeared to reduce completely over several weeks. However, within two months, all lesions had regrown. The lesions on the back regrew primarily on the edges of the treated areas. In the case of patient #4, while the lesion appeared to be slowly reducing in size at 1 month post treatment, the entire area was retreated at that time, using externally applied light from the laser (120mw/cm^2 for 10 minutes, 6 days post 2.5mg/kg Hpd). This caused rapid necrosis of the lesion which is still being followed.

Side effects of Hpd administration are primarily related to photosensitivity, resulting in marked edema and skin burn when patients accidentally expose themselves to direct sunlight either indoors or outdoors. This occurred in one of the patients who fell asleep in the sun (indoors) for 1-2 hours, 2 days following 5.0mg/kg Hpd. The only exposed portion of her body was her face, which swelled considerably within a few hours of exposure. Erythema was also evident. Ice packs were applied and swelling subsided over a week with slight sloughing of skin of the chin and lips.

Facial edema also occurred in one of the dogs exposed to sunlight in a car, shortly after receiving 5.0mg/kg Hpd. In this case, the edema subsided within three days. No other effects were noted.

During treatment, patients may experience a burning sensation near the tip of the fiber or a throbbing sensation. When this occurs, the power was reduced or, in a few cases, treatment was terminated temporarily. Because the light is delivered below the skin, little or no skin necrosis was seen, as can occur with externally applied light |1|. In one patient, areas of skin necrosis, approximately 1cm diameter, occurred over each fiber which was implanted only 5mm below the skin.

◄ Fig. 2   Patient #3 , before treatment

◄ Fig. 3   Patient #3, during interstitial photoradiation

◄ Fig. 4   Patient #3, two weeks after treatment

Discussion

The most significant finding from this study is the demonstration of erradication of osteogenic sarcoma lesions without destruction of surrounding normal tissue or bone forming elements. In the two cases where tumors had been destroyed, bone remodeling has occurred.

Clearly, the destruction from a single implant, at least as presently applied, is limited to a volume of approximately 10cc. Thus, while osteogenic sarcoma lesions of 2.5cm x 2.5cm (approximately 8cc) and 3.0cm x 3.5cm (approximately 17cc) could be eliminated with 2 or 3 implants respectively (the latter required two such treatments), a lesion 7.5cm x 6.5cm x 3.5cm (approximately 90cc) could not be controlled even with five implants. This limitation is also apparent in the patients treated to date. While in some cases transient reduction was seen, there was clearly additional tumor in these areas which was not destroyed, resulting in ultimate regrowth. Light from a single fiber implant can frequently be seen to cover an area of 3-4cm diameter on the skin surface. However, the biological effect is apparently more limited. The minimum dose of light to destroy tumor following a given dose of Hpd needs to be defined. This information, together with knowledge of light dosimetry in tissue will allow a more judicious choice of treatment conditions. One method to extend the treatment volume is to use multiple fibers, placed so as to give adequate light dose throughout an entire mass of tumor. For this application, multiple fibers can be coupled to a single laser, preferrably through beam splitting devices rather than fiber bundles, which are inherently inefficient, or multiple lasers can be used if greater intensities are necessary. One can speculate on the various potential applications of this technique. Osteogenic sarcoma is considered to be a radiation resistant tumor. The current theory for this disease is generally amputation followed by chemotherapy. If such tumors in humans respond to photoradiation therapy, it may be possible to use this therapy in combination with chemotherapy to obviate the need for amputation. Numerous other possibilities are apparent but they await extention and verification of these results.

Table 1.
Summary of results of photoradiation on primary dog and cat tumors*

*All received 5.0 mg/kg Hpd intravenously, 2 or 3 days prior to treatment.

Animal	Histology	Size of Lesion	Treatment Date	Treatment Conditions	Result
#1 Cat	Osteogenic Sarcoma/ Mandible	2.5 x 2.5cm	8/15/77	40 minutes at 300mw (inserted fiber)	10/77: X-ray negative; cat well. 6/78: No tumor; cat well; mandible reformed. 5/79: X-ray negative. 7/79: Cat healthy.
#2 Dog	Adenocarcinoma/ Pelvis	4 x 4cm	11/7/77	30 minutes at 280mw (inserted fiber)	11/77: 50% reduction; no further follow-up.
#3 Dog	Melanoma/ Mouth	Two Lesions 4 x 2cm One Pigmented	12/6/77	35 minutes at 250mw (inserted fiber)	12/77: 50% reduction in non-pigmented lesion.
			1/16/78	35 minutes at 100mw/cm² 10 minutes at 120mw/cm² (External) (Laser-multiple lesions)	2/78: 50% reduction.
		2 x 1cm Pigmented	11/4/78	20 minutes at 120mw/cm²	5/78: No tumor evident.
		0.5cm Pigmented	2/13/79	20 minutes at 300mw (inserted fiber)	12/78: 0.5cm lesion remaining. 6/79: No tumor; dog healthy.
#4 Cat	Squamous Cell Carcinoma/Mouth	2 x 1cm Pigmented	2/3/78	30 minutes at 200-320mw (inserted fiber)	2/23/78: Partial reduction.
	Squamous Cell Carcinoma/Bone	2.5cm	3/16/78	35 minutes at 130mw (inserted fiber)	3/28/78: Apparent complete clearance. 6/78: Regrowth. 7/78: Cat died.
#5 Dog	Osteogenic Sarcoma/ Tibia	7.5 x 6.5 x 3.5cm	3/17/79	20 minutes at 300mw (inserted fiber) —6 areas—	4/79: Continuing to grow.
#6 Dog	Osteogenic Sarcoma/ Maxillary Sinus, Hard Palate	3.5 x 3.0cm	6/15/79	30 minutes at 200mw (inserted fiber) —3 areas—	6/22/79: X-ray showed tumor reduced 50%.
			7/11/79	30 minutes at 400mw (inserted fiber) —3 areas—	7/27/79: No radiological evidence of tumor.
#7 Dog	Mastocytoma/ Lower Leg	2.0 x 1.0cm	8/3/79	10 minutes at 30mw/cm² (External) —3 areas—	9/1/79: Tumor sloughing; continuing follow-up.
#8 Cat	Fibrosarcoma/ Lower Leg	2.3 x 1.0cm + 0.7 x 0.5cm	8/16/79	30 minutes at 180mw (inserted fiber) —3 areas—	9/1/79: Tumor sloughing; continuing follow-up.

Table 2. Summary of results on metastatic tumors in humans

Patient Number	Type/Location/Size	Hpd Dose (mg/kg)	Interval (Days)	Laser Intensity (mw)	Treatment Time (min)	Number Fiber Placements	Result
1.	Adenocarcinoma/Labia/4x4cm	5.0	4 and 7 / 12	150 / 200 / 250	30 / 20 / 60	2 (2cm apart) / 2 / 2	Anterior portion soft without apparent mass; posterior unchanged without progression (2 months post treatment)
2.	Salivary gland carcinoma/head, neck, shoulder, chest/1.0 to 3.0cm nodules contiguous over 15-20cm	3.5	5,6	200-300	15-20	1 in each nodule, spaced 2-3cm apart	Entire area edematous, no apparent regression (1 month)
3.	Adenocarcinoma/ breast, chest wall, back, opposite breast/ 1.0 to 4.0cm contiguous lesions over wide area on back	3.5	4 and 6	300	20	2 (single) 4x2cm nodule) 7-in individual lesions on back	Complete regression (1 month) Regrowth (2 months)
			4 to 5	300	20	1 in each of 7 - 2-3cm nodules over area 10x13cm	Complete regression (1 month) Regrowth (2 months)
4.	Adenocarcinoma/ breast,chest wall/ 1-4cm	2.5	3	200-500	20	1, nodule 1cm or less 4, around periphery of 4cm nodule	<50% reduction (1 month)
5.	Adenocarcinoma/ breast, opposite breast (2/3 involvement)	5.0	5 to 8	200	60	6, around breast (3cm deep)	Follow-up incomplete.

References

1. Dougherty, T.J., Kaufman, J.E., Goldfarb, A., Weishaupt, K.R., Boyle, D.G. and Mittelman, A. Cancer Research 38:2628-2635, 1978.

2. Dougherty, T.J., Lawrence, G., Kaufman, J.E., Boyle, D.G., Weishaupt, K. and Goldfarb, A. J. Natl. Cancer Inst., 62-2:231-237, 1979.

3. Rassmussen-Taxdall, D.S., Ward, G.E. and Figge, F.H. Cancer 8:78-81, 1955.

4. Gregorie, H.B., Horger, E.O., Ward, J.L., Green, J.F., Richards, T., Robertson, H.C. and Stevenson, T.B. Ann. Surg., 167:820-827, 1968.

5. Gomer, C.J. and Dougherty, T.J. Cancer Research 39:146-151, 1979.

6. Weishaupt, K.R., Gomer, C.J. and Dougherty, T.J. Cancer Research 36:2326-2329, 1976.

7. Spikes, J. and MacKnight, M. Ann. N.Y. Acad. Sci., 171:149-162, 1970.

8. Carrano, C., Tsutsui, M. and McConnell, S. Cancer Treat. Reports 61:1297-1300, 1977.

9. Winkelman, J. Experientia 23:949-950, 1967.

10. Dougherty, T.J. Unpublished results.

11. Weishaupt, K.R. and Dougherty, T.J. Unpublished results.

Hematoporphyrin as a Sensitizer in Tumor Phototherapy: Effect of Medium Polarity on the Photosensitizing Efficiency and Role of the Administration Pathway on the Distribution in Normal and Tumor-Bearing Rats

L. Tomio[+], E. Reddi[+], G. Jori[+], P.L. Zorat, G.B. Pizzi, and F. Calzavara

Divisione di Radioterapia, Ospedale Civile di Padova and
[+]Istituto di Biologia Animale, Centro C.N.R. Emocianine,
Università di Padova
Padova, Italy

1. Introduction

The increasing number of reports on the successful treatment of different neoplasias with porphyrins and visible light (1, 2) stresses the need for a definition of the mechanism(s) responsible for the photoinduced regression of tumours, as well as for the definition of the factors enhancing the therapeutic efficiency of this novel technique.

Toward this aim, we undertook a detailed investigation of the hematoporphyrin (HMP)-sensitized photooxidation of model biological substrates in solvents of different polarity and composition; these should mimic the large variety of media which are found in "in vivo" systems. Moreover, we investigated the influence of the injected dose and of the administration pathway on the time-course of HMP accumulation in selected tissues and in tumour cells of rats affected by Yoshida ascites hepatoma. Previous studies in our laboratory (3) showed that HMP exhibits a high affinity for ascitic tumours.

2. Solvent effect on the HMP-sensitized photooxidation of L-tryptophan

The aromatic amino acid L-tryptophan typically undergoes photosensitized oxidation by a mixture of type I and type II mechanisms (4). The former photoprocess involves attack by photoexcited HMP (in its lowest triplet state) on the indole ring with the generation of radical intermediates, e.g. through electron transfer; the latter photoprocess involves energy transfer from triplet HMP to molecular oxygen leading to the formation of an activated oxygen derivative, singlet oxygen (5). Therefore, this amino acid is a suitable substrate for studying the solvent effect on the relative weight of the two aforesaid photosensitization mechanisms.

Typically, 10 ml of a 0.1 mM tryptophan solution containing 0.083 mM HMP was exposed to the light of four 250 W tungsten lamps. As a solvent, different water (buffered at pH 7.4)-methanol and –formamide mixtures were used. The dielectric constants of methanol and formamide at 20°C are 32.63 and 109, respectively, as compared with a value of 78 for water. The solutions were air–equilibrated and were maintained at 17°C by circulating water.

In all the reaction media investigated, the light–induced decrease of tryptophan concentration, as monitored by spectrophotofluorimetric analysis (6), followed first–order kinetics. The photooxidation rate constants in the different solvent mixtures were calculated from the slope of the semilogarithmic plots and were subsequently plotted versus the percentage of organic solvent (Fig. 1). The bell–shaped plot, obtained upon increasing the methanol concentration, suggests that a change in the photoreaction mechanism takes place at about 30 percent methanol. Now, both the solubility of oxygen and the lifetime (hence, the reactivity) of singlet oxygen (7) are greater in methanolic than in aqueous solution; therefore, it appears reasonable to hypothesize that the ascending portion of the plot reflects a more and more important contribution of the singlet oxygen photoreaction mechanism. The type I (radical–involving) pathway should become predominant above 30 percent methanol; actually, in this case, the radical intermediates are endowed with a greater polarity than the original reactants, hence their formation should be more difficult when the medium polarity is lowered. This may explain the steady decrease of the photoreaction rate observed

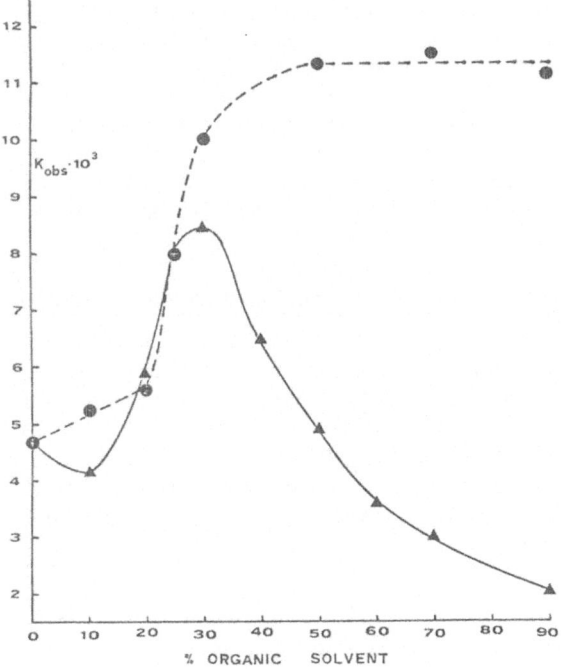

Fig.1 Effect of the concentration of organic solvent on the first-order rate constant of the HMP-sensitized photooxidation of tryptophan. The solvent was made by mixing suitable volumes of water (pH 7.4) with methanol (triangles) and with formamide (circles).

77

at high percentages of methanol. The shift of the tryptophan photooxidation mechanism from a type II to a type I when the organic solvent concentration increases is confirmed by the following
evidences:

i) the addition of sodium azide (a specific quencher of singlet
   oxygen) to the irradiated system results in a strong inhibition of the photoprocess up to 30 percent methanol; above this
   methanol concentration the inhibitory power of azide becomes
   weaker and weaker in agreement with a reduced efficiency of
   the singlet oxygen pathway.

ii) the HMP-sensitized photooxidation of other substrates, which
    proceed exclusively by attack of singlet oxygen (e.g., histi-
    dine or methionine at a concentration of 0.1 mM), is characte-
    rized by a steady enhancement of the photoreaction rate going
    from water to 90 percent methanol.

iii) performing the photooxidation of tryptophan in water-forma-
     mide mixtures, i.e. in media of increasing polarity, causes
     a gradual increase of the photoreaction rate up to about 40
     percent organic solvent, with no detectable subsequent lower-
     ing of the rate (Fig. 1). Clearly, the radical intermediates
     are stabilized by the greater polarity of the medium.

Now, spectroscopic studies on HMP in the different reaction
media described above unequivocally demonstrate that the porphyrin exists in an aggregated state in the 0-30 percent organic
solvent concentration range. Above this value, HMP undergoes a
sharp transition to a monomeric form. Therefore, our findings
strongly suggest that aggregated HMP performs its photosensitizing action mainly by energy transfer to singlet oxygen; on the
other hand, the monomeric dye, in its electronically excited
triplet state, preferentially interacts with tryptophan to initiate radical processes. Therefore, the efficiency and the mechanism of the HMP-photosensitized reactions are controlled by
both the polarity of the medium and the concentration of the dye
(which, in turn, affects the equilibrium between the monomeric
and the aggregated form).

3. Time-course of HMP distribution in normal and tumour-bearing rats

In normal rats after both intraperitoneal (i.p.) and intravenous
(i.v.) injection of HMP, out of the various organs examined (liver, kidneys, spleen and muscle), the greatest amount of dye was
accumulated in the liver. The uptake of HMP by the liver was a
remarkably rapid process. The accumulation was kinetically independent of the administration pathway, although a larger amount
of HMP was present in the liver following i.p. administration
(Fig. 2).

In tumour-bearing rats, the liver again displayed the highest
affinity for HMP. In the case of i.p. injection, HMP diffusion

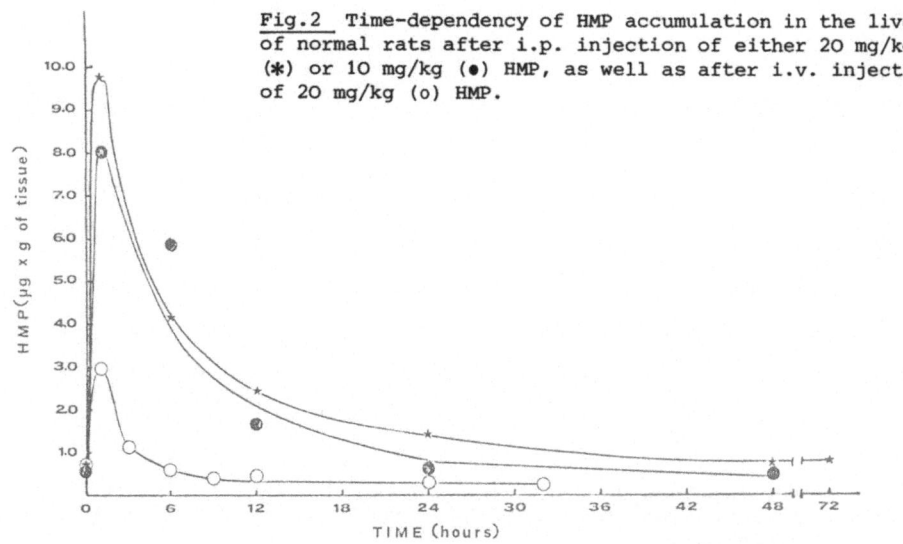

Fig.2 Time-dependency of HMP accumulation in the liver of normal rats after i.p. injection of either 20 mg/kg (*) or 10 mg/kg (●) HMP, as well as after i.v. injection of 20 mg/kg (o) HMP.

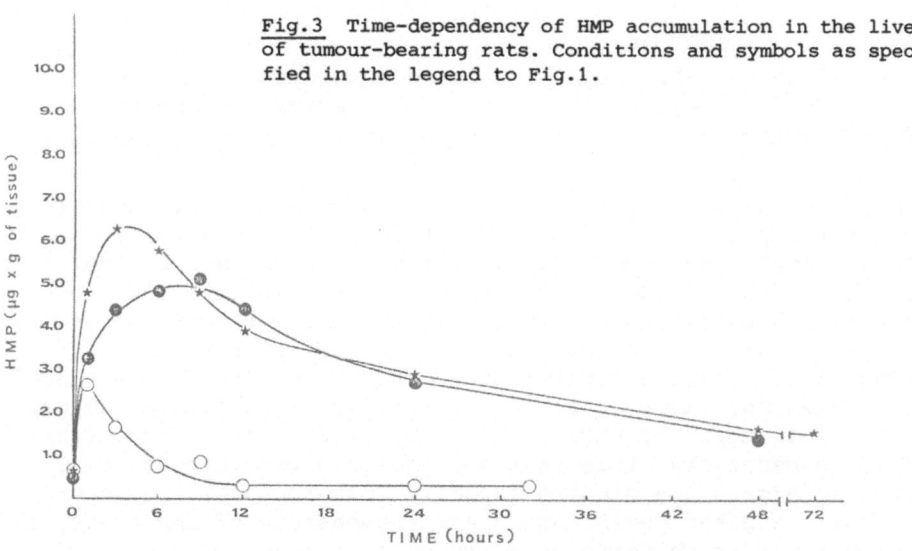

Fig.3 Time-dependency of HMP accumulation in the liver of tumour-bearing rats. Conditions and symbols as specified in the legend to Fig.1.

to the liver was slower (maximal accumulation at 3 h, as one can see from Fig. 3), probably owing to the presence of the ascitic tumour; the latter, however, had no appreciable effect on the kinetics of HMP distribution after i.v. administration. Analogously, the dose of i.p.-injected HMP affected the time-course of HMP distribution among the various organs, including liver, of tumour-bearing rats, while it was ineffective for normal rats, as one can deduce from a comparative analysis of figs. 2 and 3.

79

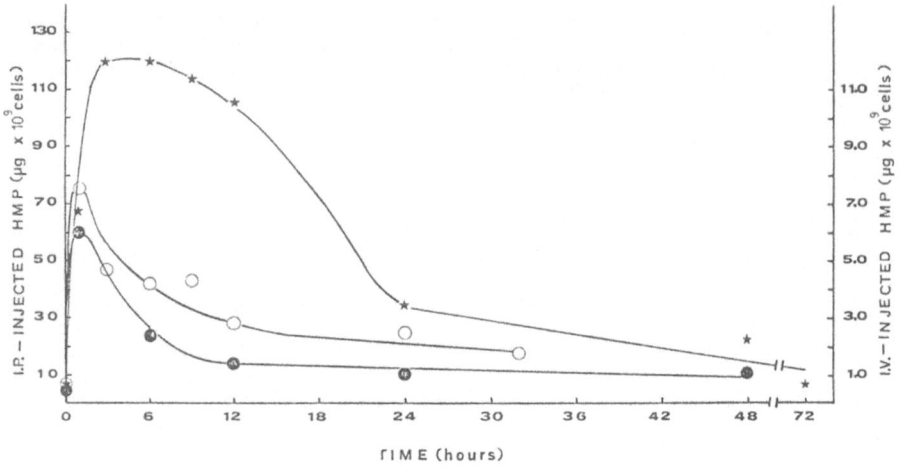

Fig.4 Time-dependency of HMP accumulation in ascitic tumour cells after
either i.p. or i.v. administration of HMP. Conditions and symbols as specified in the legend to Fig.1.

    In all cases, HMP preferentially accumulated in the neopla-
stic cells (Fig. 4); the HMP uptake was maximal at about 1-3 h
after injection and remained at relatively high levels up to
9-12 h.

    Our data provide a clear demonstration that HMP exhibits a
high affinity also toward tumour cells growing in suspension,
such as ascites hepatoma. The selectivity of HMP uptake from
neoplastic tissues is further shown by the time-dependency of
the ratio between the HMP concentration in the tumour and in the
liver (see Table 1). Apparently, the tumour/liver ratio is dependent both on the administration mode and on the dose of injected HMP. For example, at 12 h after i.p. injection of 20 mg/
kg and of 10 mg/kg of HMP, the tumour/liver ratio was about 27
and 3, respectively. This fact may indicate that a dose of 10
mg/kg can exceed the binding capacity by serum proteins, such
as albumin and hemopexin -which are responsible of HMP transport
(8) - but not the binding capacity by tumour cells.

    Our data are in agreement with the findings of WINKELMAN and
RASMUSSEN-TAXDAL (9), who studied the HMP affinity for a carcinoma in rats. However, some differences are found when our data
are compared with those obtained by DOUGHERTY et al. (10) upon
studying the distribution of a HMP derivative in mice affected
by a transplanted mammary carcinoma. Therefore, it is necessary
to verify the conditions leading to maximal pigment accumulation
for each porphyrin-tumour system.

Table 1. Time-dependency of the ratio between HMP concentration in the tumor cells and in the liver of rats affected by Yoshida ascites hepatoma

Hours	Injection mode		
	e.v.	i.p. (10 mg)	i.p. (20 mg)
1	2.84	18.17	14.05
3	2.97		19.15
6	5.44	4.92	20.76
9	6.13		23.42
12	9.32	3.22	26.93
24	7.81	3.99	11.39
32	6.10		
48		9.23	8.59

4. Conclusions

1) The photosensitizing efficiency and mechanism of HMP depends on the aggregation state of the dye, hence on the injected dose; high HMP doses should favour the singlet oxygen pathway, so that cellular structures far from the site of HMP binding may be readily inactivated; on the other hand, low HMP doses enhance the probability of radical mechanisms, which occur within a narrow spatial range.

2) In rats affected by ascitic tumours, the distribution pattern and tissue concentration of HMP depends on dose, administration pathway and time. A critical factor is represented by the HMP binding capacity of serum proteins, which are responsible for HMP transport to the liver.

3) Until now, phototherapy has been mainly applied to the treatment of cutaneous lesions (both metastatic and recurrent), where the ratio between porphyrin accumulated in the tumour and in surrounding skin is very high (11). Our data open the way to the phototreatment also of deep-sited tumours, provided the suitable porphyrin dose and administration pathway are chosen and some technical problems are solved.

5. References

1. Granelli S., Diamond A., McDonaugh A., Wilson C. and Nielsen S., 1975, Cancer Res. 35, 2567-2570.
2. Weishaupt K.R., Gomer C.J. and Dougherty T.J., 1976, Cancer Res.,36, 2326-2330.

3. Jori G., Pizzi G., Reddi E., Tomio L., Salvato B., Zorat P.L. and Calzavara F., 1979, Tumori 65, 425-434.
4. Cannistraro S., Jori G. and Van de Vorst A., 1978, Photochem. Photobiol. 27, 517-521.
5. Sconfienza C., Van de Vorst A. and Jori G., 1979, Photochem. Photobiol., in the press.
6. Genov N. and Jori G., 1973, Int. J. Peptide Protein Res. 5, 127-133.
7. Foote C.S., 1976, in "Free Radicals in Biology" (W.A. Pryor, ed.) Vol. II, pp. 105-133, Academic Press, New York.
8. Koskelo P.,and Muller-Eberhard U,, 1977, Seminars Hematol. 14, 221-226.
9. Winkelman J. and Rasmussen-Taxdal D.S., 1960, Bull. John Hopkins Hosp. 107, 228-233.
10. Dougherty T.J., Grindey G.B., Fiel R., Weishaupt K.R. and Boyle D.G., 1975, J. Natl, Cancer Inst. 55, 115-129.
11. Dougherty T.J., Kaufman J.E., Goldfarb A., Weishaupt K.R. and Mittleman D.A., 1978, Cancer Res. 38, 2628-2635.

Acknowledgment. This work was supported by the Consiglio Nazionale delle Ricerche (Italy), contract no. 79.00607.96, under the progetto finalizzato "Controllo della Crescita Neoplastica".

Spectroscopic Properties of Human Skin and Photodestruction of Tumors

A. Anders[1] and P. Aufmuth[2]

Institut für Biophysik, Universität Hannover[1], Institut A für Experimentalphysik
Universität Hannover[2]
D-3000 Hannover 1, Fed. Rep. of Germany

E.-M. Böttger and H. Tronnier

Hautklinik der Städtischen Kliniken
D-4600 Dortmund 1, Fed. Rep. of Germany

1. Introduction

In the first part of this paper we report on laser investigations of absorption and transmission of human skin, untreated and treated with photosensitizing dyes. In the second part we study the application of dye laser irradiation in the phototherapy of tumors.

In connection with phototherapy and photochemotherapy, spectroscopic properties of human skin are of practical interest. E.g., a deeper understanding of the penetration of light into tissue, localisation of the dye in cells, or the reason for selective incorporation of some dyes in tumor cells are important questions. The use of tunable lasers as light sources for photochemotherapy offers, besides high spectral intensity, the advantage of wavelength tunability. Thus, one can excite with a narrow band in the maximum of the action spectrum of the photosensitizer [1,2]. The total irradiation intensity and possible unwanted reactions may be reduced in this way.

In the phototherapy of tumors, laser irradiation in the visible and UV region is discussed. Until now, the irradiation has been performed with visible light, utilizing the photodynamic action of the dye on cells as a destruction process [3,4]. Irradiation in UV may become the basis for a new kind of photodestruction [2]. When the irradiation takes place below 310 nm - where the dye as well as the nucleic acids absorb - the nucleic acids transfer energy to the dye molecules [5]. This interaction between nucleic acids and dye may cause destruction of the tumor.

2. Spectroscopic Properties of Human Skin

We used samples of isolated human epidermis from amputations. The skin specimens were mounted between quartz plates [6].

Continuous wave dye lasers, an Ar^+ laser and a pulsed dye laser were applied. The pulsed dye laser served as test laser to measure the transmission and reflexion of the skin. The continuous wave dye lasers and the Ar^+ laser were used as irradiation sources; before and after irradiation, the transmission was measured with the test laser [6].

With dye lasers a far better resolution in absorption and transmission spectra of human skin specimens could be obtained than with spectrophotometers [6]. Changes of transmission occur in skin (untreated with dye) depending on wavelength and intensity of the irradiation and the test wavelength [1]. Which molecular reactions are responsible for such processes has to be decided in further experiments.

Skin treated with thiopyronine or acridine orange shows an increase in transmission after irradiation which is caused by a change of the dye to its leukoform [1,7]. The correlation between the transfomation of the dye to the leukoform and the photodynamic effect can yield information about details of the photodynamic process, e.g., the maximum of the action spectrum of the dye. During the bleaching of the dye free radicals can be formed which may interact with cell molecules.

3. Phototherapy of Basal Cell Carcinomas with Dye Lasers

In the phototherapy of tumors we have studied, up to now, six tumors with two patients. Basal cell carcinomas were stained with acridine orange. We applied externally a 0.01 % solution of acridine orange in ethanol and painted it on the tumors (some ml per cm^2 in several steps). After drying they were irradiated with a flashlamp-pumped dye laser (Chromatix CMX-4) [2] at the wavelengths λ = 490 nm and λ = 296 nm. The energy density amounted to 6 J/cm^2 (490 nm) and 0.02 J/cm^2 (296 nm). At 296 nm, the energy density was much smaller because of the greater absorption of skin in the UV region (see Fig. 1.c). Additionally, only stained and only irradiated basaliomas were compared. All basal cell carcinomas were excised two days later, and frozen cell sections were studied with a fluorescence microscope. Table 1 summarizes the main results.

With visible irradiation mainly subepidermal incorporation in tumor cells was found; little fluorescence was observed in epidermis layers, except in regions of ulcerations. UV irradiation leads to a strong fluorescence in basal epidermis layers

Table 1   Staining and irradiation of basal cell carcinomas

Staining: acridine orange
Dye laser irradiation: λ = 296 nm, 490 nm

Only staining:	no fluorescence
Only irradiation:	no effect
Staining and irradiation with 296 nm:	fluorescence especially in tumor cells
Staining and irradiation with 490 nm:	subepidermal fluorescence, especially in tumor cells

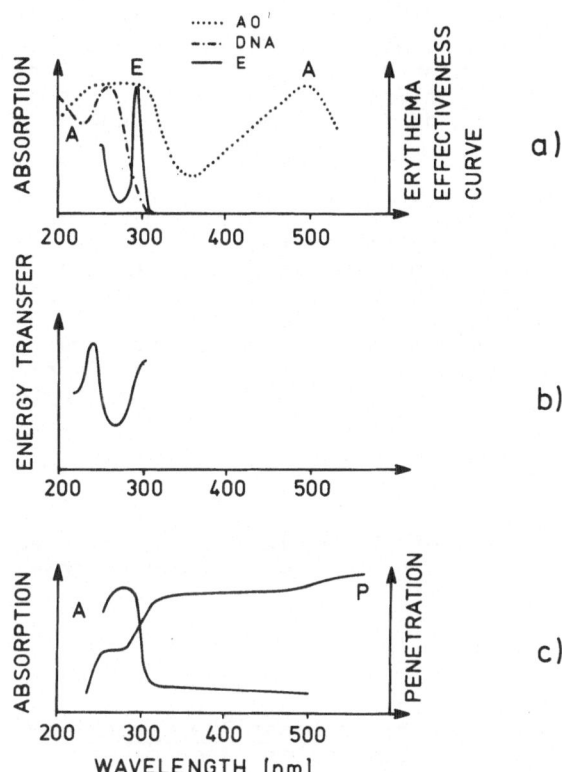

Fig.1 (a) DNA absorption and acridine orange (AO) absorption in skin and standard erythema effectiveness curve; (b) typical energy transfer from DNA to acridine orange [5]; (c) absorption of epidermis (A) and penetration of light into the cutis (P). All curves are normalized to a maximum value of 1

and, in regions of ulcerations, to fluorescence of the connective tissue, too. The observed fluorescence suggests binding of the dye in these cells. In mice tumor cells, TOMSON [3] found binding of acridine orange to the cell nuclei.

The aim of our preliminary studies with patients was the incorporation of acridine orange after pulsed laser irradiation in the visible and UV. Further experiments are to be related to the observation of cell destruction with increased total irradiation intensity.

4. Discussion: UV Irradiation in the Phototherapy of Tumors

Irradiation in UV should lead to an interaction of the photosensitizer with the nucleic acids as in the photochemotherapy of psoriasis. Figure 1 summarizes the facts which have to be considered in the discussion of this problem.

Irradiation below about 310 nm excites the dye as well as the nucleic acids (Fig. 1.a). Interaction of the dye with the nucleic acids implies a binding of acridine orange to the cell nuclei, such a binding was found in mice tumor cells [3]. Acridine orange interacts with DNA by intercalation; in complexes of this kind energy transfer processes increase near the absorption

85

edge of the nucleic acids at 300 nm (Fig. 1.b) [5]. An investigation has to be made as to whether this interaction between dye and nucleic acid may cause tumor destruction, alone or additionally to the photodynamic process. An interesting question, too, is the quantity and mode of action of free radicals produced under UV irradiation.

Around 300 nm the applicable irradiation intensity is limited because of the erythema reaction. To avoid this limitation the irradiation wavelength may be shifted to shorter wavelengths where the erythema curve decreases. On the other hand, this entails a decrease in energy transfer (Fig. 1.b). In the UV region a further disadvantage is the reduced penetration of light into tissue (Fig. 1.c); the value is about 30 % at 280 nm and about 40 to 45 % around 300 nm in comparison to 55 % at 500 nm. Overall it should be possible to find a compromise between all these effects.

Acknowledgement

We gratefully acknowledge the help of Chromatix GmbH for placing a laser CMX-4 and equipment at our disposal for the tumor irradiation.

References

1. A. Anders, H. Zacharias, P. Aufmuth: In Laser 77 - Opto-Electronics, ed. by W. Waidelich (IPC Science and Technology Press, Guildford 1977) pp. 520-526
2. A. Anders, P. Aufmuth, E.-M. Böttger, H. Tronnier: In Laser 79 - Opto-Electronics, ed. by W. Waidelich (IPC Science and Technology Press, Guildford 1979) pp. 355-360
3. S. Tomson: Optics Laser Technol. 81-84 (1976)
4. T.J. Dougherty, J.E. Kaufman, A. Goldfarb, K.R. Weishaupt, D. Boyle, A. Mittleman: Cancer Res. 38, 2628-2635 (1978)
5. A. Anders: Opt. Commun. 26, 339-342 (1978)
6. A. Anders, I. Lamprecht, H. Schaefer, H. Zacharias: Arch. Derm. Res. 255, 211-214 (1976)
7. A. Anders, H. Zacharias, I. Lamprecht, H. Schaefer: In Medizinische Physik, Vol. 2, ed. by W.J. Lorenz (Hüthig Verlag, Heidelberg 1977) pp. 539-547

Photodynamic Effect of Hematoporphyrin (HP) On Cells Cultivated in Vitro

T. Christensen and J. Moan

Norsk Hydro's Institute for Cancer Research, Montebello
Oslo 3, Norway

1. Introduction

Recent clinical trials with photochemotherapy of human tumors have been successful [1, 2]. Many of the mechanisms leading to destruction of tumors after treatment with hematoporphyrin derivatives and light are yet unidentified. Experiments with cells in vitro have been useful in the search for cellular effects of photochemotherapy. There are, however, situations in treatment of tumors that are difficult to simulate in vitro.

The delivery of light to the entire tumor mass has been a problem in the treatment of solid tumors. To overcome this difficulty, the use of laser-fiber optics has been useful [3]. However, even with multiple fibers embedded in the tumor mass, it must be expected that the light intensity will be variable throughout the tumor. This will probably lead to different degrees of damage in different sites of the tumor. Factors like variable binding of dye to cells in well and less well vascularized areas in the tumor may also lead to different degrees of damage. Contact between cells in different areas may modify the response towards photodynamic destruction.

In this paper we report an effect of contact between cells during treatment with HP and light.

2. Materials and Methods

Cells from the established line NHIK 3025 were used in this study. The line is derived from a carcinoma in situ and is cultured in medium E2a containing 20% human serum and 10% horse serum. Subcultivation is done 3 times every week and the cells are in almost continuous exponential growth. The mean duration of the cell cycle is 17 - 18 h ($G_1 \sim$ 6.5 h, S $\sim$ 8 h, $G_2 \sim$ 2.5 h and M $\sim$ 1 h).

The experimental procedure is shown in Fig. 1. A cell population is treated with 0.25% trypsin (Difco 1:250) and inoculated as single cells in 25 cm^2 Falcon tissue culture flasks. After a variable period the cells received $4 \cdot 10^{-4}$ M HP (Sigma) in fresh medium. HP was kept in contact with the cells for 30 min at 37° C prior to irradiation and was also present during light treatment. This was performed by irradiating the attached cells through the bottom of the flasks with light from two "black-light" lamps (Osram) at 365 nm. The light intensity reaching the cells (5 cm above the lamps) was measured to 11.0 w/m^2 with a calibrated thermopile (Yellow Springs Instruments). After irradiation the HP-containing medium was removed and the cells were allowed to form macroscopic colonies in fresh medium. These colonies were stained and counted after about a week incubation. Cells able to form macroscopic colonies were scored as surviving.

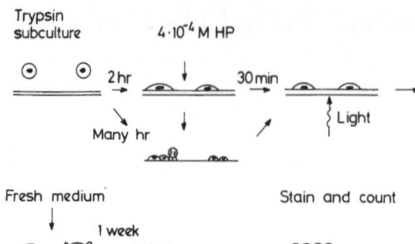

Fig.1 Experimental method used to determine the sensitivity of attached NHIK 3025 cells towards photodynamic inactivation.

Treatment with HP and light followed variable periods after the inoculation of single cells. During this period the cells continued dividing and formed microcolonies. In order to follow the multiplication, growth curves were made by scoring mean multiplicity in at least 50 microcolonies.

In one experiment the effect of 2 min treatment with 0.25% trypsin was studied. This treatment was done with microcolonies and was sufficient to detach the colonies from the substratum. Care was taken not to separate the cells and the colonies were allowed to attach for 2 h before treatment with HP and light.

3. Results and Discussion

A dose response curve for single cells is shown in Fig. 2. It has a continuous downward bend with a flat part at low light doses. Populations of asynchronous cells consist of subpopulations with different sensitivities towards photodynamic inactivation[4]. The flat part of the dose response curve can be explained by the relatively greater contribution to the sensitivity from the most sensitive cells at low light doses. Likewise the curve shape at high doses will be characteristic for the most resistant subpopulation.

The time between trypsinization and treatment with HP and light has been varied in Fig. 3. The growth curve shows that the number of cells per colony increases exponentially. If the sensitivity of each cell had been independent of colony size one should expect the curve showing sensitivity as a function of time after trypsinization to become almost parallel to the growth curve at low survival levels. This is obviously not the case in the two experiments shown in Fig. 3. On the contrary, colonies of 2 and 8 cells have about the same sensitivity as single cells. The small variations at times shorter than 10 h seen in one experiment is believed to be due to synchronization effects after trypsin treatment.

The variation in sensitivity as a function of colony size is investigated further in Fig. 4. Each panel shows two dose response curves made for the same cell population inoculated at different times. Panel A shows that the sensitivity up to 6 h after trypsinization is constant. The dose response curves for microcolonies (open symbols) in panels B and C show a little lower sensitivity at low doses and a little higher sensitivity at higher doses for microcolonies compared to single cells.

It can be concluded that the sensitivity of microcolonies with up to 8 cells is not less than that of single cells at light exposures longer than about 15 min. The higher "shoulder" on the dose response curves for microcolonies can be interpreted as a result of cell kinetics. Effects of cells

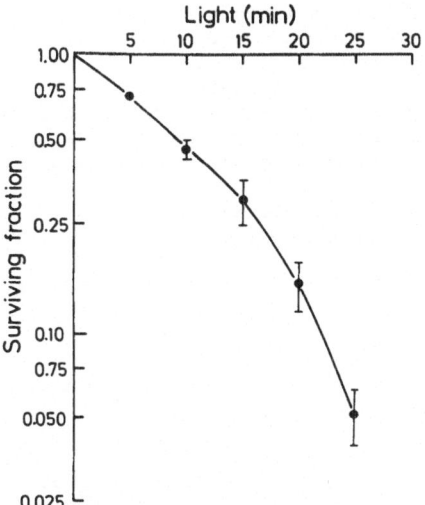

Fig.2 Sensitivity of single NHIK 3025 cells towards photodynamic inactivation in the presence of $4 \cdot 10^{-4}$ M HP exposed to 11.0 W/m^2 of light at 365 nm. Mean and S.E. from 6 experiments.

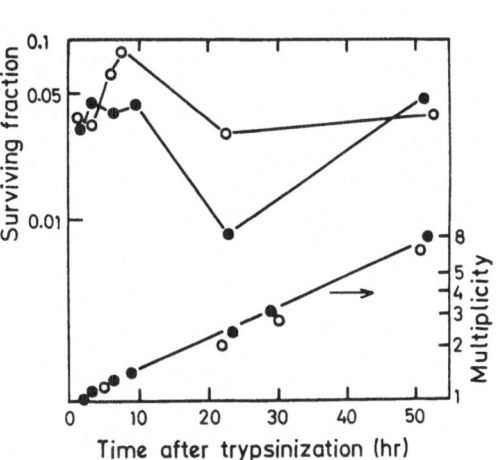

Fig.3 Upper curve: Survival of microcolonies after treatment with HP and 20 min light at different times after trypsin treatment. Results from two experiments. Lower curve: Corresponding mean multiplicity of microcolonies.

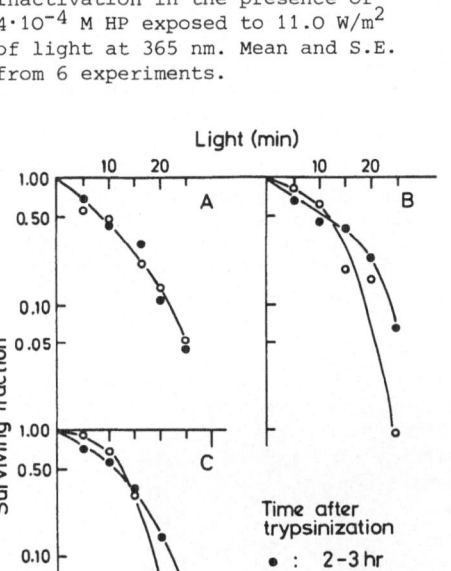

Time after
trypsinization
● : 2–3 hr
○ : A 6 hr
    B 30 hr
    C 46 ½ hr

Fig.4 Dose response curves for cells treated at different times after trypsin treatment. Multiplicity: 2–3 hr 1.0 – 1.15, 6hr 1.46, 20 hr 2.79, 46½ hr 4.58.

in the particularly sensitive S phase will probably be concealed by more resistant cells in the same colony.

A possible explanation is that the trypsin treatment during subcultivation makes the cells more resistant and that they are becoming progressively more

sensitive as time elapses after the trypsin treatment. This is checked by
subjecting microcolonies to an additional trypsin treatment. A set of flasks
were inoculated with cells as described and incubated for 44 hr. Then the
microcolonies were given a mild treatment with trypsin sufficient to detach
the colonies from the substratum. After a period of re-attachment of about
2 hr, the colonies were exposed to HP and light. These colonies had nearly
as high multiciplity as colonies cultivated similarly except for the additional trypsin treatment (3.95 vs. 4.58). It is seen from Fig. 5 that this
additional treatment with trypsin does not make the microcolonies more
resistant than single cells at light doses higher than 15 min.

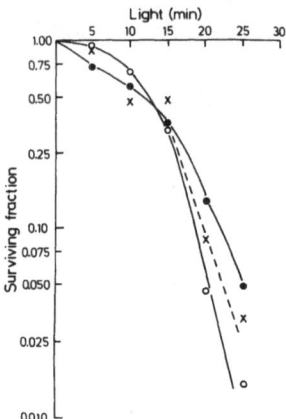

Fig.5 The effect of additional trypsin treatment of colonies. The symbols are: Closed
circles, cells exposed to HP and light 2-3 hr
after subcultivation (multiplicity 1-1.15).
Open circles, microcolonies exposed to HP and
light 46½ hr after subcultivation (multiplicity
4.58). Crosses and broken line, microcolonies
given treatment with trypsin 44 hr after subcultivation and exposed to HP and light 46½ hr
after subcultivation (multiplicity 3.95).

These results leave one possible explanation, there must be a co-operation
between cells, either in the processes leading to inactivation or in the
binding of HP. The former is favoured by direct microscopic observation
of microcolonies treated with HP and light. Typically one finds two types
of appearance, either colonies where all cells are lysed or colonies where
all cells appear unaffected by the treatment (Fig. 6). In such experiments
it is seldom one finds colonies where both visibly damaged and non-damaged
cells are present.

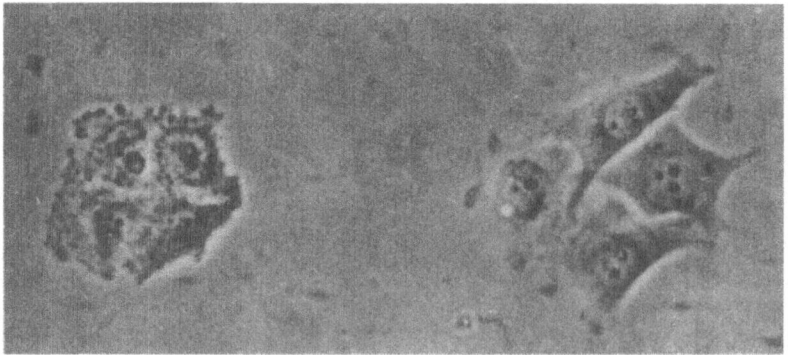

Fig. 6 Two colonies from the same flask treated with HP and light. The
picture is taken 1 h after light treatment.

These results indicate that the co-operation between cells can both protect a cell from damage and transfer damage to it. The nature of this
mechanism is still unclear, but it is tempting to speculate if transport
of toxic photoproducts may be involved. Photodynamic inactivation of cells
is postulated to proceed via singlet O_2. This molecule diffuses only a
distance about 0.1 µm in cellular environments [5, 6] . The transfer of
singlet O_2 between cells is therefore not likely.

After photodynamic treatment leakage of lysosomal-like enzymes presumably from the lysosomes [7] and accumulation of water inside the cells
[6] have been observed. Inter-cellular transport may lead to transfer of
these products between cells.

A collective killing of cells is expected to be of great importance if
it takes place in tumors. Such effects may lead to destruction of cells
in areas of the tumor not accessible to high intensities of light for geometric reasons.

References

1. J.F. Kelly, M.E. Snell, J. Urol. 115, 150 (1976)

2. T.J. Dougherty, this book.

3. T.J. Dougherty, D.G. Boyle, K.R. Weishaupt, D. Yakar, R. Thoma,
    Proc. Am. Ass. Cancer Res. 20, 420 (1979)

4. T. Christensen, J. Moan, E. Wibe, R. Oftebro, Br. J. Cancer 39, 64 (1979)

5. K.R. Weishaupt, C.J. Gomer, T.J. Dougherty, Cancer Res. 36, 2326 (1976)

6. J. Moan, E.O. Pettersen, T. Christensen, Br. J. Cancer 39, 398 (1979)

7. A.C. Allison, I.A. Magnus, M.R. Young, Nature 209, 874 (1966)

This study was supported by The Norwegian Cancer Society.

Laser Fluorescence Bronchoscopy for Early Lung Cancer Localization

D.R. Doiron and A.E. Profio

Department of Chemical and Nuclear Engineering, University of California
Santa Barbara, CA 93106, USA

1. Introduction

Lung cancer is responsible for over 89,000 deaths annually in the United
States. In order to decrease this rate, early diagnosis, localization, and
therapy must be undertaken. Early detection is available by sputum cytology.
An early lesion may only be a few millimeters in extent and 100 micrometers
thick. Such a small preinvasive lesion is not localizable by conventional
radiography, computed tomography, or nuclear medicine. The lesion may be
visually indistinguishable from normal bronchial mucosa variations under white
light fiberoptic bronchoscopy.

Previously we described a fluorescence bronchoscope system for localization
of small lung tumors using the tumor-specific marker, hematoporphyrin derivative (HpD), and a filtered mercury arc lamp excitation source [1]. The marker
is injected intravenously at a dosage of 1.5 to 2.5 mg/kg body weight three
days prior to the fluorescence bronchoscopy procedure. The tumor is imaged
by its fluorescence while the surrounding normal tissue is dimmer in comparison. Corrected fluorescence excitation and emission spectra for HpD in human serum are shown in Fig. 1. This instrument detected larger lung tumors
[2]. Difficulty was found in distinguishing thin lesions because of reflected
red background from the excitation source.

A laser excitation source has replaced the mercury arc lamp in order to
allow localization of smaller and thinner tumors. It gives the added advantage of higher delivered excitation illumination with a lower red background
compared to an arc source. These improvements have been found to be highly
advantageous in visualizing thin lesions.

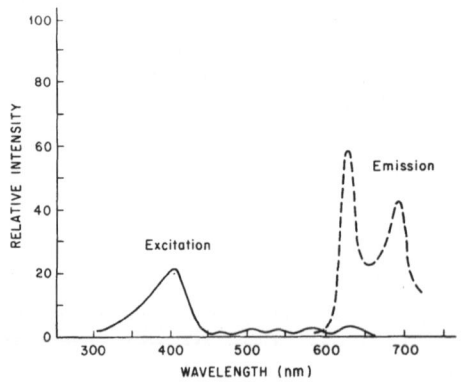

Fig.1 HpD fluorescence excitation
and emission spectra in human
serum at 27°C

2. Laser Fluorescence Bronchoscope

Figure 2 is a schematic diagram of the laser fluorescence bronchoscope system.

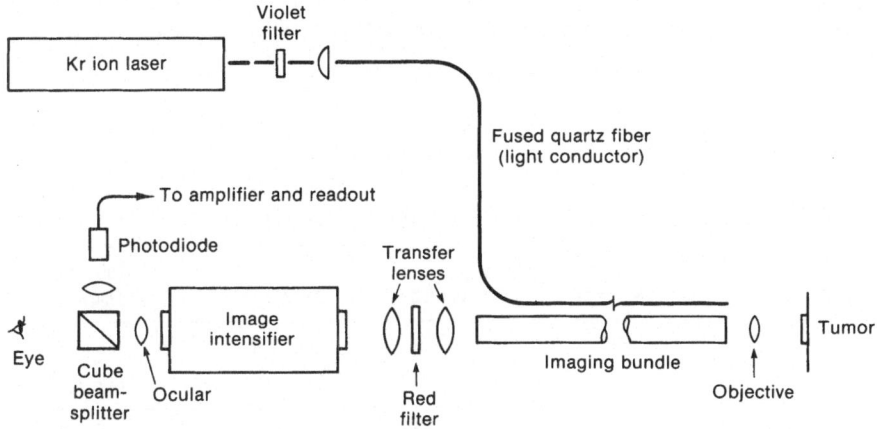

Fig.2 Laser fluorescence bronchoscope system schematic

Fluorescence excitation is provided by a CW krypton ion laser set up to lase in the violet at 406.7nm (36%), 413.1nm (60%), and 415.4nm (4%). These wavelengths lie well within the Soret absorption band of the HpD. The laser is a Spectra Physics Model 164-11/265 with special mirrors and a high field magnet, giving a total typical output of 240mw. The excitation is delivered to the bronchial mucosa via a single 400μm core quartz fiber inserted in the biopsy channel of a fiberbronchoscope. The fluorescence yield of a small tumor is on the order of 10^{-5}, hence a low background is essential. The red discharge of the laser is blocked by a filter to give a red to violet power ratio of 10^{-6} out of the fiber. By comparison the filtered mercury lamp ratio is 10^{-5}.

A single fiber system eliminates the need for a special enhanced violet-transmitting bronchoscope as required with an arc source. It also frees the normal illumination bundle for easy switching to white light examination. One difficulty with a quartz fiber is the small numerical aperture which causes only a portion of the field of view to be illuminated. Solutions to this problem are being investigated. However, the bronchoscopist has learned to compensate for the restricted illumination.

Fluorescence luminance out of the imaging bundle is on the order of 10^{-3} $\ell m/cm^2$ for a small, thin lesion. With such a small signal, it is necessary to intensify the image for visual recognition. This is done by using a three-stage, electrostatic-focus intensifier (Varo 8858). A red barrier filter is positioned between the imaging bundle and the intensifier photocathode to reject reflected excitation light and any other nonred background such as autofluorescence of cartilage. The image intensifier has a S20-extended red photocathode and a P20 phosphor output screen. It converts the faint red fluorescence image to a bright green one with a typical brightness gain of 50,000. This green image matches the high acutence photopic vision.

3. Results

Small lesion phantoms consisting of 100μm deep slots filled with solutions of varying HpD concentration have shown the system's ability to detect an HpD mass density of 10pg/cm^2. This corresponds to a total mass of 0.71pg in the design target lesion (70μm thick by 2.0mm in diameter). It is capable of imaging a 2:1 contrast ratio for an HpD concentration of 1.0μg/ml.

At this time, two patients with bronchial tumors have been examined using the laser system. The results are summarized in Table 1.

Table 1 Laser fluorescence bronchoscopy patient results

Patient	Chest X-ray	Bronchoscopy White	Fluorescent	Biopsy[a]	HpD Dose (mg/kg)
E.V.	+	+	+	Ad. CA.	2.5
J.H.	-	+	+	Sq. CA.	2.0

(a) Ad.= Adeno, CA.= Carcinoma, Sq.= Squamos

Patient J.H. showed no visible sign of a lesion on chest x-rays, but showed a positive malignancy by sputum cytology. The lesion was only localized by white light bronchoscopy after two examinations. Fig. 3 is a photograph of the fluorescing lesion in the apical segment of the right upper lobe. Serial sectioning after surgical resection showed the lesion to be only 1mm thick at its greatest depth, with areas of carcinoma-in-situ.

These trials were done at the University of Alberta Hospital, Edmonton, Alberta by Dr. E. G. King and his associates. Future trials are planned as suitable patients become available.

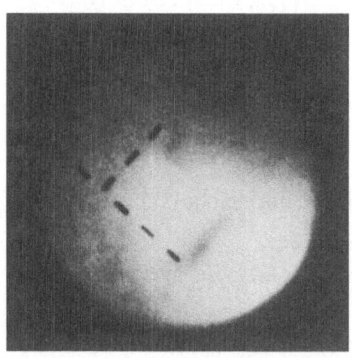

Fig.3 Photograph of occult bronchogenic tumor in apical segment of the right upper lobe. Dotted tee is inscribed on the window of the intensifier

4. Violet Phototherapy

Dougherty has been successful in treating cancer patients with HpD and red light at 630±10nm [3]. Activation of the photodynamic reaction is an HpD absorption dependent process. HpD absorption at the krypton laser lines is approximately twenty times greater than at 630nm. Preliminary cell culture experiments with a xenon lamp filtered for 410±5nm has shown that at a dose rate of 1mw/cm^2, the violet light is some 70 times more effective than at 630±10nm. At lower dose rates, a repair mechanism appears to be present. These results are being investigated further at this time.

Violet light is absorbed strongly by hemoglobin and biological pigments. For this reason, its use in photodynamic therapy is limited to small, thin lesions and is unsuitable for administration through the skin. The efficient coupling of a laser to a single fiber, however, allows for insertion into the cancerous tissue via a hypodermic needle. Work is planned to investigate this concept. Penetration would be sufficient in a thin bronchial tumor.

5. Acknowledgments

This research was supported by the Division of Biomedical and Environmental Research, U.S. Department of Energy, under Contract EY-76-s-03-0113 with the Medical Imaging Science Group, University of Southern California, School of Medicine. The laser was loaned by Spectra Physics, Inc. We acknowledge the collaboration of E. G. King, M.D., Thomas J. Dougherty, Ph.D., and Gerald C. Huth, Ph.D.

6. References

1. A. Edward Profio and Daniel R. Doiron, "Phys. Med. Biol." 22, 949 (1977)

2. Daniel R. Doiron, A. Edward Profio, Ronald G. Vincent, and Thomas J. Dougherty, "CHEST" 76, 27 (1979)

3. Thomas J. Dougherty, et al, "Cancer Research", 38, 2628-2635, 1978

Part III

Photodermatology

The Future Prospects for Lasers in Dermatological Photobiology

I.A. Magnus

Department of Photobiology, Institute of Dermatology, Homerton Grove
London E9 6BX, England

The purpose of this paper is to give first an historical review of conventional light sources as used in photodermatology, secondly to note their
advantages and disadvantages, and lastly to make suggestions as to where lasers
may be useful. Photodermatology uses artificial light for two purposes; either
for treatment or for investigations.

Artificial light sources in photodermatology may be dated back to Niels
FINSEN [1] who first developed the carbon arc for therapy. FINSEN was a medical man by training; but he tackled his subject like a physicist and
engineer. He had to make his own lamps and optical equipment and find out
for himself what part of the spectrum was useful for the purpose in hand. He
found lenses were necessary to focus the radiant energy and he had to evolve
filters for removing heat and visible light, e.g. ammoniacal copper sulphate,
a fore-runner of Wood's glass. He successfully treated Lupus vulgaris, a
mutilating skin disease very common in the late 19th century that had no
treatment of benefit, and FINSEN's carbon arc therapy was soon adopted world
wide. Its raison d'etre is less clear. Was it due to a germicidal effect?
FINSEN seems to have favoured this and used bacteria as a means of investigating both action spectra and lamp dosimetry. Or was the cure of Lupus vulgaris
from photosynthesis in the skin of Vitamin D? That this could be so follows
from the work of DOWLING in London and CHARPY in Paris, who, independently in
the 1940s found oral calciferol (given without skin irradiation) cured the
disease. But, as Lupus vulgaris is a form of tuberculosis of the skin, as
soon as effective antibiotics that killed the tubercle bacillus became readily
available in the 1950s, these new drugs were used instead. As a result
phototherapy fell into almost complete disuse and, where still used, even
came into some disrepute.

The only other well established beneficial effect of UV irradiation of the
skin, apart from FINSEN's treatment, was the production of Vitamin D. UVR
could therefore be used to treat diseases due to Vitamin D lack such as in
rickets, a condition in children resulting in bone deformity, and in osteomalacia, the same disease in adults. The adult condition may occur in pregnant
women, but also in the elderly sick or geriatric patients. The latter are a
serious problem because such patients tend to be bedridden and indoors for
long periods of time and, as well, very often suffer dietary deficiencies.
These ills are all corrected by UVR therapy. But again vitamin D deficiency
can be cured very simply by oral preparations of the vitamin, so UV irradiation
therapy was again soon outmoded in this sphere as well as in the treatment of
Lupus vulgaris. However, there may be a small application for preventive
phototherapy in geriatric hospital patients, especially where there is difficulty in patients taking a normal diet.

Phototherapy nevertheless has lingered on over the last 50 years with a
rather discredited reputation. It had been overenthusiastically used for
a number of conditions in the first three or four decades of this century
during a time when critical methods for appraising the results of therapy
were not well established. Medical men were in a small way in a kind of
dilemma, torn between, on the one hand, apparent beneficial effects of impressive irradiation procedures (which the patients enjoyed) and, on the other
hand, the application of the laws of common sense and the philosophy of therapeutic nihilism. In fact the only other clearly beneficial effects of UV
phototherapy was in dermatology for the treatment of Psoriasis and Acne
vulgaris.

We have to admit, however, that this view is contested in the USSR and
other Eastern European countries where, following the school of research of
Danzig and his colleagues, UVR had found application in preventive medicine
in the field of public health. Small exposures of UVR, given fairly frequently, are thought to prevent minor illnesses in and improve the performance of
school children and factory workers. It is claimed that the giving of small
doses of UVR similarly may play a beneficial and preventive role to all those
living in a more-or-less windowless or sunless environment; it is thus suggested that small doses of UVR should be emitted by all light sources used
indoors.

Only very recently has a resurrection occurred in phototherapy of skin
diseases. This is with so-called PUVA therapy, the use of psoralen (furocoumarin) given orally or locally together with irradiation of the skin with
a predominantly near-UVR source. There is no doubt that PUVA is popular with
both patients and many doctors. Its long term safety, however, remains
controversial and obscure and whether it really offers any advantage over
treatments already available is not yet clear.

Before leaving photobiology we may note in passing the cult of skin UV
irradiation with artificial light in the demi-monde, the down town sauna
or red light area massage parlour, to promote tanning of the skin. Equally
remarkable though perhaps less disreputable is the cult of the acquisition
by solar exposure of the suntanned skin as a meritorious symbol of "one-
upmanship." Only sociologists will be able to explain why a brown skin is
at present held in such favour; this was not always so, for in the Old
Testament and in the Renaissance European literature, it was the pale skin
which was held in esteem, see for example instances in the plays of William
SHAKESPEARE.

I now pass on to other uses of conventional light sources. Conventional
light sources have made very great contributions to photodermatology in their
time. FINSEN's share of this has already been noted in his development of
the air burning carbon arc and his application of UVR to therapy. This spurred
others to develop other lamps, in particular the mercury vapour arc, which
in its medium pressure version with a number of discrete emission lines in
the UV spectral region around 250-300 nm, lent itself to use as a source for
a monochromator, in the pioneering work of HAUSSER [2] and his colleagues in
the early 1920s and the discovery of the action spectrum for sunburn and
suntan. The sterling work of SHULZE and his colleagues [3] followed in the
same field using the same medium pressure mercury arc.

Another important conventional source is the low pressure mercury vapour
arc. Being monochromatic at 254 nm, its use has been most fruitful in micro-

biology where photobiologists have played most important roles in studying damage to DNA. WACKER and his colleagues [4] in Germany first discovered the fact that 254 nm radiation causes a photolesion in living bacterial DNA, the thymine dimer. This was followed by SETLOW and his colleagues [5] and by HOWARD-FLANDERS and his co-worker [6] in 1964 simultaneously and independently describing a repair system for the UVR induced thymine dimers in bacterial DNA. This is the so-called "excision repair" system. HOWARD-FLANDERS and his workers [7] then discovered a second bacterial DNA lesion repair process, so-called "post replication repair." These two DNA repair processes occur in human tissue and the marked reduction of one or other of them is a feature of the disease known as Xeroderma pigmentosum. This disease is associated with extreme skin sensitivity to UVR and to the development in childhood of malignant tumours leading to premature death. James CLEAVER [8] first found the skin of patients with this disease to have greatly reduced excision repair of 254 nm UVR induced thymine dimers; Alan LEHMANNN and his colleagues [9] showed that in other patients with this disease it was the post replication DNA repair system that was at fault, not the excision repair. All this work was based on the use of the simple 254 nm emitting low pressure mercury vapour arc, and, as the work on Xeroderma pigmentosum is one of the first instances where one can plausibly suggest a possible "cause of cancer" at a molecular level, it should be regarded as a very important light source where irradiance need not be high. The 254 nm low pressure mercury lamp also led to the development of fluorescent tubes and a large assortment of lamps of different spectral outputs, these have been of use mostly in therapy. Time does not allow considering them much here.

The high pressure xenon arc as a source with a very strong continuum has also played a useful role in photodermatology. The compact arc xenon lamp proved suitable for investigative work with high intensity monochromators and such instruments have clarified the role of light and UVR in the abnormal photosensitivity in the metabolic disorders known as the porphyrias. Xenon arc powered irradiation monochromators are also very useful for studying skin photosensitivity in the group of conditions known as "photosensitve eczema" and "actinic reticuloid." In the latter conditions and in the porphyrias, abnormal skin photosensitivity lies in the near UVR and visible spectrum and rather large doses of light may sometimes be required for testing patients thought to have these disorders. Hence very high dose rates are desirable.

Here we come upon a difficulty, when using an irradiation monochromator, of deciding how to choose between a narrow waveband which is spectrally pure but of relatively low irradiance, or a wider waveband which is of relatively high overall irradiance. With the classical monochromator, which works by dispersing the spectrum, one cannot have both good spectral purity and high power. A compromise must be arrived at. Generally in photodermatology we choose the wide waveband, rather spectrally impure, of high irradiance. This is for the obvious practical reason that patients cannot be immobilized for long periods; usually 15 minutes or so is sufficient for a test exposure of the skin in an adult; much less time, say 5 minutes, is enough for a child.

Another particularly difficult problem in the design and use of high intensity irradiation monochromators is keeping scattered or stray light reduced to as low a value as possible. Where a xenon arc is used as a monochromator's source, scattered light is mostly from the visible spectrum as the chief output from the lamp is visible and infra red; this is perhaps less of an objection if one is using a narrow waveband in the UVR spectrum, which is biologically very much more active than visible. However, using say a narrow waveband in the near UVR, one is particulary concerned about stray radiation in the shorter UVR. Giving a dose of 50 Joules per square metre of a narrow waveband centred at 360 nm may, let us say, entail a dose of 50 millijoules per square metre scattered shorter UVR of 300 nm and wavelengths shorter. The latter would probably be sufficient to say a false positive reaction. The same considerations apply with the use of the super high pressure mercury vapour arc. The answer to these problems, of course, is to use a laser.

My suggestions for the performance requirements of lasers for use in photodermatology are as follows: 1) Monochromaticity: This is one of the substantial advantages of lasers in certain types of biological work and it should be preserved as far as possible. 2) Collimation: Again it is very important to preserve this remarkable feature of the laser. 3) Wavelength range: 200-800 nm. 4) Exposure rate: This is not known in all applications, but watts to kilowatts per square metre are probably desirable. 5) Exposure dose: 10 kilojoules to 1 megajoule per square metre. 6) Target area: From 1 mm^2 to about 1 m^2. 7) Pulsed or continuous emission : Which is ideal is not known, but probably continuous emission would be preferable. 8) Coherency: Relevance unknown. These requirements are not all available, but no doubt they will be in the near future. Lasers offer the important features of excellent monochromaticity with, in many cases, no fears as to the possible adulteration by stray light. The laser monochromatic wavelength, is, of course, very often monochromatic in the strict sense and not just a more or less narrow waveband. Possible applications of lasers in photomedicine would be: 1) To measure skin reflectance. 2) To perform skin irradiation tests. 3) For skin therapy: (a) for the individual treatment of skin lesions by 300 nm UVR in Acne vulgaris (b) localized fields - individual treatment of small psoriatic lesions in PUVA-type treatment (c) generalized or large field - extensive lesions in (a) and (b). 4) For application in: (a) preventive medicine, e.g. so-called "hygienic" uses as per in the USSR in windowless environment or in arctic territories (b) internal diseases: (i) osteomalacia (ii) icterus neonatorum (iii) photodynamic treatment e.g. of internal cancer, such as in endoscopic treatment of bladder cancer.

References

1. N.R. Finsen: Om Anvendelse I Medicinen af Koncentrerede Kemiske Lysstraaler (Gyldendalske Boghandels Forlag (F. Hegel & Son), Copenhagen 1896) pp. 5-52
2. K.W. Hausser and W. Vahle: Stralentherapie 13, 41-71 (1922)
3. U. Henschke and R. Schulze: Strahlentherapie 64, 14 (1939)
4. A. Wacker, H. Dellweg and D. Weinblum: Naturwissenschaften 20, 477 (1960)
5. R.B. Setlow and W.L. Carrier: Proc. Nat. Acad. Sci. 51, 226-231 (1964)
6. R.P. Boyce and P. Howard-Flanders: Proc. Nat. Acad. Sci. 51, 293-300 (1964)
7. W.D. Ruff and P. Howard-Flanders: J. Mol. Biol. 31, 291-304 (1968)
8. J.E. Cleaver: Nature 218, 652-656 (1968)
9. A.R. Lehmann, S. Kirk-Bell, C.F. Arlett, M.C. Paterson, P.H.M. Lohman, E.A. de Weerd-Lastelein and D. Bootsma: Proc. Nat. Acad. Sci. 72, 219-223 (1975)

Will Lasers be Useful in Photodermatology?

J.C. van der Leun

State University of Utrecht, Institute of Dermatology
Utrecht, The Netherlands

1. Introduction

Photodermatology has grown vigorously during the past two decades.
An increasing number of clinics and institutes has joined work
in this field and the trend is still continuing. In the context
of a conference on a special kind of light source it may be good
to notice that this growth was not primarily due to the availability of new types of equipment. Much of the recent work could
also have been performed with apparatus available in the previous
heyday of skin photobiology, in the years around 1930.
  In my assessment, the recent growth of photodermatology was
stimulated more by a few pressing needs. In the clinics it became
increasingly clear that something had to be done about photodermatoses. Suffering from sunlight is a bad thing, and the number of patients having such complaints is increasing. Interest
was further stimulated by the recognition of the fact that sunlight was the cause of most skin cancers, and also the main
cause of aging of the skin. The latest stimulus was provided by
the possibility that stratospheric ozone would be reduced by
pollution; this would lead to increasing intensities of solar
ultraviolet radiation.
  It is not my intention to belittle the rôle of equipment,
but at a conference such as this it may be good to realize that
equipment has not been the driving force in the recent past, and
may not be so in the future. Nevertheless, improved equipment
of course plays an important rôle. It may facilitate further
developments and even extend the limits of existing observational
methods. With the field of photodermatology in accelerating
growth, the advent of the practicable ultraviolet laser appears
well-timed. I will try to assess its possible contributions from
a few different viewpoints.

2. Economy

Patients having photodermatoses are being "photo-tested". This
starts with exposure of small fields of the patient's skin to
light of a number of wavelengths, each in several doses. This
is very time-consuming, especially if one wishes to use many
wavelengths. There is little doubt that a tunable UV laser of
sufficient intensity will markedly reduce the time required for
such diagnostic work. At least in the UV-C and UV-B, heat puts
no limit on the shortness of the exposures. Reduction of the

time required for the exposures will make the diagnostic facilities available to more patients. This in turn may well help to increase our knowledge; data collected in this type of diagnostic work have formed the basis of much of our knowledge in skin photobiology [1].

The economy of intense monochromatic light sources is potentially even more striking in phototherapy, where total-body exposures are common.

Let me take the "PUVA"-therapy as an example. In this therapy, patients with various skin diseases are exposed to UV-A after having been photosensitized, usually by 8-methoxypsoralen. Radiant doses usually range between 1 and 20 J/cm^2 per exposure. The exposures are given in light-cabinets; fluorescent UV-A lamps with the highest intensity achievable are assembled in closest packing on the walls. All this effort is only just sufficient to administer the doses required in acceptable exposure times; these still may run up to 30 minutes. During this time the patient is under a heavy heat load, although the irradiance in the relevant wavelength region is not more than 10 mW/cm^2, an order of magnitude less than the total irradiance in summer sunlight. It is obvious that monochromatic light would be a real improvement. Heating of the total body would hardly be a problem. Even the largest doses used at present could not raise the body temperature by as much as 1^0 centigrade. If the output of the laser would pose no limit, the UV-A could be administered at a rate only limited by the heating of the skin. This would lead to exposure times shorter than 2 minutes, with less discomfort to the patient than in the present facilities.

The gain to be achieved in phototherapy with UV-B is even greater. The doses of UV-B given are in the order of magnitude of the "minimal erythema dose" (MED). The MED of unexposed skin approximately equals 20 mJ/cm^2. Such a dose can be given to the skin without heat damage in one flash, no matter how short the laser engineers can make it. Even if the sensitivity of the patient would decrease by a factor of 100 by the repeated exposures during the treatment, the dose required could still be administered without heat problems in about 15 seconds.

Shortening of exposure times is not only convenient to patients and staff, it may also help in reducing the cost of treatments.

3. Specific laser effects

One might hope to gain new insights from investigations utilizing the special characteristics of laser radiation. The coherence of the radiation or the concentration of energy in short pulses might produce special effects, giving new information or opening new investigative possibilities.

As far as skin photobiology is concerned, we are only in a very early stage; a few privileged investigators have the opportunity of beginning to probe the possibilities of lasers. The first reports [2],[3] show results not drastically deviating from what would be expected on the basis of extrapolation. One is inclined to be struck rather by the degree of correspondence with earlier results. I am thinking especially of the provisional confirmation of reciprocity for the erythemal effect of UV-A [2];

this means that the effect of the radiation only depends on the dose, the product of irradiance and exposure time. Reciprocity appears to hold even when the radiant dose is administered in pulses much shorter than ever have been available.

It is too early to tell whether or not specific laser effects will be found in skin photobiology, and whether or not these will open new insights. I have no intention at this moment to go into science fiction.

4.   Extension of observational data

Lasers have in any case the advantages of monochromaticity and intensity. A degree of monochromaticity satisfactory to skin photobiology may well be achieved with lamps and monochromators, but it is almost always compromised by the requirement of sufficient irradiance. Many interesting experiments had to be stopped at a certain point for lack of irradiation facilities. Any skin photobiologist will be able to give examples. I will mention a few from my personal experience.

In studying ultraviolet erythema, the reddening of the skin caused by ultraviolet radiation, early investigators noticed that the effectiveness of the radiation in eliciting erythema was strongly dependent on wavelength (Fig.1).

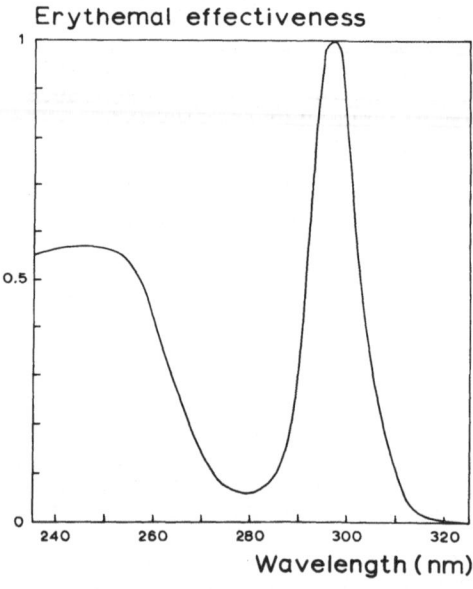

Erythemal effectiveness

Wavelength (nm)

Fig.1   Action spectrum of ultraviolet erythema (CIE, 1935)

Figure 1 shows the classic action spectrum, recommended in 1935 by the Commission Internationale de l'Eclairage. Many action spectra have been determined before and after that time, and all show curves of different shapes [4]. Almost all agree in a few

general aspects: that erythemal effectiveness is high in the
wavelength regions around 250 nm and 300 nm, and that there is
a more or less developed minimum in between.

In studying the erythemas elicited with wavelengths around
250 nm and around 300 nm, some investigators have been impressed
by the differences between the effects of the two wavelength
bands [5],[6]. This led to the explicit hypothesis that there
are two distinct erythemal mechanisms, one with an action spectrum extending through the entire wavelength region from 250 nm
to 400 nm and one limited to a narrow wavelength region around
300 nm [6]. The hypothesis is illustrated in Fig.2.

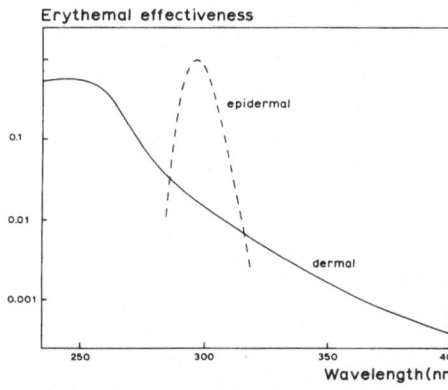

Fig.2 Hypothesis of the composite action spectrum

These two mechanisms would each fit one of the long-standing
ideas about the genesis of the erythema. The observable reddening
of the skin is caused by increased blood content, by dilation of
the small superficial blood vessels in the dermis. One idea is
that the ultraviolet radiation acts directly on these blood
vessels, or at least in the level of these blood vessels, and so
causes the dilation. The other idea is more complicated. The
radiation acts on the more superficial skin layer, the epidermis;
there it causes the liberation of a mediator substance which diffuses to the blood vessels in the dermis, where it effects the
dilation. The broad action spectrum would fit the idea of direct
action on the dermis, the narrow peak around 300 nm would represent action on the epidermis, the diffusion idea.

There were of course indications in favour of this hypothesis,
but also indications putting question-marks. The dispute has not
yet been settled conclusively, and one of the reasons is lack of
observational possibilities.

Erythemas elicited with wavelengths around 275 nm could give
a clue, but these have hardly been studied. And in the few cases
that investigators tried it there is reason for doubting whether
they did not in fact study effects of straylight coming from the
bordering wavelength regions with higher erythemal effectiveness.

The UV-A has for a long time been the Cinderella of skin
photobiology, in this case not for lack of monochromaticity, but
for lack of irradiance. This again has not facilitated conclusions.

There is usually some time between the exposure of the skin
to ultraviolet radiation and the onset of reddening of the skin.

This interval is called the "latent time". The higher the dose of UV radiation, the shorter the latent time is. Examination of the relationship between the dose of UV and the latent time gives a possibility of distinguishing between various erythemal mechanisms. The diffusion conception gives a specific relation, which may be compared with observations. It is represented by the curve in Fig.3.

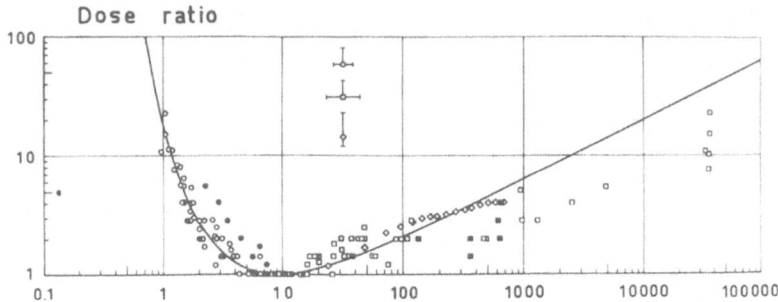

Fig.3 Dose-time relationships of ultraviolet erythema. The dose ratio is the dose expressed as the number of MED's. The curve is given by a theory based on the diffusion conception. The points denote observations on the 300-nm erythema; open circles represent the appearance of such erythemas on the forearm, closed circles on the thigh. Squares represent disappearance of erytemas (not discussed here).

The figure shows that the fit is quite good for 300-nm erythemas. Similar observations on 250-nm erythemas, not shown in the figure, produced a far less close agreement with the diffusion curve. The obvious desire, to perform a similar investigation in the UV-A, could not be fulfilled. The minimal erythema dose in the UV-A is so high that with conventional equipment it takes an exposure time of about half an hour to elicit a just perceptible erythema. For the investigation discussed one would need various multiples of this exposure time, and soon the exposure time would exceed the latent time to be measured. Using a laser giving 337-nm radiation, PARRISH and co-workers could produce perceptible erythemas with a mean exposure time of 3.5 minutes [2]. With such a laser the investigation discussed could be extended into the UV-A.

Another attempt to distinguish between the two erythemal mechanisms was made by investigating the lateral extension of the redness beyond the boundaries of the exposed skin area. Such extensions are indeed measurable; a sample of measurements is shown in Fig.4.

At the onset of the erythema, it is smaller than the exposed skin area. It increases in size, reaches a maximal extension and then retreats again. The discontinuity in the curve is related to the peeling of the skin.

106

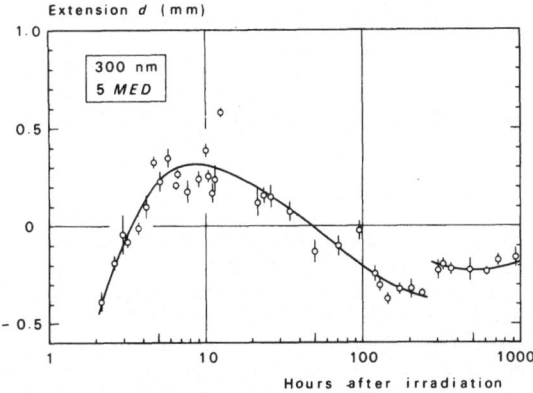

Fig.4 Extension of a 300-nm erythema beyond a boundary of the exposed skin area. Positive values of the extension mean that the erythema is larger than the exposed skin area.

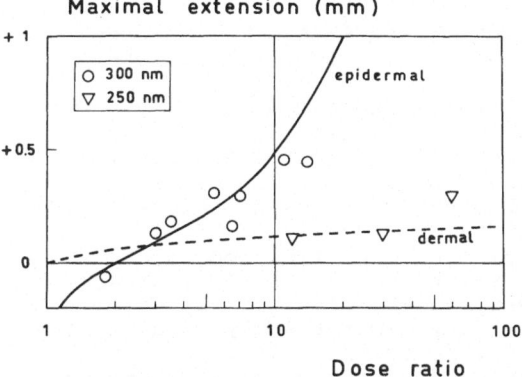

Fig.5 Relationship of the dose of UV radiation given and the maximal extension of the resulting erythema beyond a boundary of the exposed skin area. The points represent measurements, the curves two different theories.

The maximal extension achieved was dependent on the dose of UV radiation given. The relationship of maximal extension and dose was investigated for erythemas elicited with wavelengths around 250 nm and around 300 nm. The measurements are shown in Fig.5. The curves are theoretical relationships derived on the basis of the two different erythemal mechanisms.

Again it is clear that the observations on the 300-nm erythema are in reasonable agreement with the diffusion conception; the 250-nm observations are not. This is one of many observational results behind the hypothesis illustrated in Fig.2,[7]. One wonders how the UV-A erythema would fit into this picture. At the time of these measurements, I did not see a chance to arrange the required uniform exposure of 5 cm^2 of skin to UV-A in

107

sufficient doses. With present lasers the measurement is possible and I would be quite curious to know the outcome.

These are just a few examples to indicate where I would have
use for UV lasers. Even from my own work I could give several
more, and I am sure that this is similar for other skin photobiologists.

5. Conclusions

It seems clear that UV lasers will be very useful in photodermatology. Right now there are realistic possibilities to extend our observations and to resolve a few long-standing problems.
Lasers also appear to have the potential for improving the economy of several hospital procedures. This may well be helpful
in promoting that UV lasers will soon be used in many of our
institutes and clinics.

References

1. Magnus IA: Dermatological Photobiology, Blackwell Sc Publ,
   Oxford, 1976.
2. Parrish JA, Anderson RR, Ying CY and Pathak MA: Cutaneous
   effects of pulsed nitrogen gas laser irradiation. J invest
   Dermatol 67:603,1976.
3. Anders A, Lamprecht I, Schaefer H and Zacharias H: The use
   of dye lasers for spectroscopic investigations and photodynamic therapy of human skin. Arch dermatol Res 255:211,1976.
4. Johnson BE, Daniels F,Jr and Magnus IA: Response of human skin
   to ultraviolet light. In: Photophysiology, Ed. Giese AC, Vol 4,
   Academic Press, New York, 1968, pp 139-202.
5. Rottier PB: Ultraviolet radiation and skin; some facts and
   some problems. Proceedings First Int Photobiological Congress,
   Amsterdam 1954, Veenman, Wageningen 1954, pp 192-204.
6. Van der Leun JC: Ultraviolet Erythema; a study on diffusion
   processes in human skin. Thesis Utrecht, 1966.
7. Van der Leun JC: On the action spectrum of ultraviolet erythema. In: Res Progress in Organic, Biological and Medicinal
   Chemistry, Vol 3, Part II, Eds. Gallo U and Santamaria L,
   North Holland Publ Cy, Amsterdam, 1972, pp 711-736.

A Survey of the Acute Effects of UV Lasers on Human and Animal Skin

R.R. Anderson and J.A. Parrish

Department of Dermatology, Harvard Medical School, Massachusetts General Hospital
Boston, MA 02114, USA

Abstract

Cutaneous responses to nitrogen (337.1 nm), KrF (248 nm) excimer, and ArF (193 nm) excimer lasers were studied in guinea pigs and humans. In most respects, the erythema, pigmentation and histologic responses to these nanosecond-domain pulsed lasers are identical with responses to conventional CW sources within each spectral region. The N_2 laser was used to quantitatively document and compare the enhanced phototoxic responses induced by oral and topical psoralens, and various topically applied tar preparations. For nitrogen and KrF lasers, reciprocity between irradiance and exposure duration is observed for the production of the clinically observed threshold for delayed erythema responses. This is not the case for induction of immediate erythema or immediate pigment darkening by N_2 laser. The minimal erythema dose to ArF excimer laser (193 nm) was greater than 0.5 J/cm^2 in albino guinea pigs. The insensitivity to 193 nm radiation as compared with 248 nm radiation is presumably due to the higher optical density of the stratum corneum and epidermis for wavelengths shorter than 200 nm. Research applications of pulsed UV laser systems in dermatological photobiology are briefly discussed.

Introduction

The advent of pulsed lasers with high peak and average power outputs in the ultraviolet region of the spectrum offers the photobiologist a unique tool for studying responses to extremely high photon densities and very short exposure durations. In addition, the excellent monochromaticity and high average powers available make some lasers superior to conventional sources of narrow band UV radiation.

Responses to UV radiation at the clinical, tissue, and cellular levels are complex biologic reactions to photoproducts formed by the irradiation. In general, for single-photon photochemical reactions yielding relatively stable photoproducts, the yield of photoproducts is determined by the number of photons absorbed times the quantum yield for a given reaction. The quantum yield is typically a constant for a given wavelength and set of environmental conditions. Because the number of photons absorbed is proportional to the exposure dose, one often finds that the yield of photoproducts, and hence some defined photobiologic response, is determined by the exposure dose, and not the exposure duration or irradiance per se. That is, a given exposure dose at a given wavelength produces a given degree of response independent of irradiance or exposure duration. This condition is known as reciprocity. Using flash lamps, CLAESSON et al., [1] reported reciprocity to hold for leakage of dye from vessels in UV-irradiated mouse skin at peak broadband UV irradiances up to 10^3 W/cm^2. However, one would expect reciprocity failure to occur at higher peak irradiances.

In general, significant two-photon absorption may occur for power densities greater than approximately 10^6 W/cm^2, and often leads to a gross change in the quantum yield or nature of photochemical reactions within biologic tissues. The probability for two-photon absorption is proportional to the square of the peak irradiance and hence causes deviations from reciprocity. RUBIN has reported, for instance, that the quantum yield of thymine dimers induced in DNA and the inactivation of Candida guilliermondii at 265 nm is markedly enhanced for irradiances greater than 10^7 W/cm^2 [2]. Other reasons for deviations from reciprocity in photobiologic responses include biologically significant heating of the tissue during exposure, induction of active repair processes during exposure, or responses to some threshold concentration of highly reactive secondary photoproducts, whose concentrations are generally irradiance-dependent. Another special case of reciprocity failure involves photoproducts which can be formed only by sequential photoreactions separated by some period of time. The cross-linking photoproducts of some psoralens with double stranded DNA are thought to proceed by formation of a single strand photoproduct, a period of conformational relaxation, and then a second photoreaction with the opposite strand. The formation of monoadducts show first-order dependence upon exposure dose, while the formation of cross-links shows second-order dependence. HEARST et al., have shown that a single 10-nanosecond 347 nm laser pulse produces only the single strand photoproduct, while two pulses separated by more than a few microseconds produce a statistical number of crosslinks [3]. Cellular repair of these two different lesions seems to follow two different mechanisms [4]. We are reporting here on the nature of clinical and histologic cutaneous responses to pulsed UV lasers in both humans and guinea pigs, on attempts to observe reciprocity failure in response of human skin exposed to pulsed UV lasers, and on some practical uses of these lasers as sources for experimentation.

Nitrogen Laser: The nitrogen gas laser used (Avco Model C5000) produces 10 ns duration, 1 mJ pulses at 337.1 nm, with a peak power of 10^5 W and a pulse rate up to 500 Hz. The maximum average irradiance studied was 106 mW/cm^2, over a 1.8 cm^2 area. By focussing the beam and scanning it uniformly over 25 cm^2 areas of skin, peak irradiances up to 2 x 10^6 W/cm^2 were obtained while the average irradiance was only 7.5 mW/cm^2, over sites large enough to record clinical erythema and pigmentation responses (Fig. 1). Biopsies were obtained 24 and 48 hr after exposure in both humans and guinea pigs for comparison of histologic effects of this near-UV laser with a conventional near-UV source (broadband 320-400 nm radiation from a xenon arc source). In all, 25

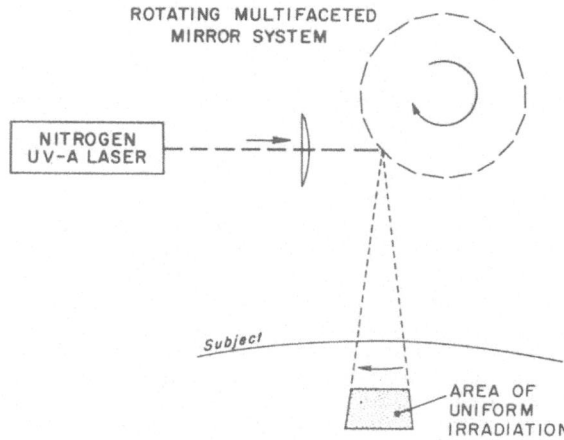

ROTATING MULTIFACETED
MIRROR SYSTEM

NITROGEN
UV-A LASER

Subject

AREA OF
UNIFORM
IRRADIATION

Fig.1. Fair-skinned
Caucasoid subjects were
studied.

In Caucasian skin, average irradiances of approximately 60
mW/cm^2 or greater can produce an erythema (redness) observable
immediately after exposure. Average irradiances less than 40
mW/cm^2 did not induce this response. At 100 mW/cm^2, approximately
2.5 min is required to produce immediate erythema, while at 70
mW/cm^2, approximately 10 min is required. The lack of reciprocity
and the immediate presence of erythema are suggestive of a ther-
mally-induced erythema. Measurements of skin surface temperature
during exposures to various average irradiances revealed that the
response was observed only for exposure durations producing a
surface temperature of 42°C or greater, regardless of the average
irradiance chosen. This immediate erythema response, and its
deviation from reciprocity, are apparently due to heating of the
tissue. Our results here are analogous to those obtained by
BUCKER [5], using conventional sources.

Another immediate response noted clinically was that of immediate pigment darkening (IPD), which is a result of photooxidation
of preformed melanin [6]. The degree of IPD was variable and
apparently dependent upon the pigmentation of the subjects tested.
IPD response to the N$_2$ laser was consistent with that observed
for conventional sources of UVA. The IPD diminishes over a period
of minutes to hours, depending upon the exposure dose and subject.

Delayed erythema response appeared clinically similar to that
produced by conventional UVA sources. Maximum delayed erythema
response occurred 18 to 24 hours following exposure. Depending
upon the severity and duration of the immediate erythema, delayed
erythema sometimes blended with it. The two erythemal responses
were essentially independent responses in the sense that delayed
erythema can be produced without immediate erythema, and vice
versa, depending upon the average irradiance used. Unlike for
immediate erythema, the minimal dose causing delayed erythema
(MED) was not a function of average irradiance. Reciprocity for
delayed erythema was studied by focussing and scanning of the beam
to produce a 200-fold increase in peak irradiance over that of the
unfocussed beam. This procedure did not affect the MED for delay-

ed erythema. Attenuating the peak irradiance to 10^3 W/cm^2 also
did not affect the delayed erythema MED. Thus, over a 10^4 range
of peak irradiance, reciprocity was observed for delayed erythema.
An apparently normal delayed tanning response was also induced in
human skin, beginning 3 to 5 days following nitrogen laser exposure. The minimal tanning doses were determined by observations
at 7 days after laser exposures, and were again consistent with
those for conventional sources in this spectral region.

Analogous experiments were performed after topical application
of 1.0% 8-methoxypsoralen (8-MOP) or 4,5',8-trimethylpsoralen
(TMP) in ETOH, applied at 32 µl/cm^2, and occluded for 1 hr before
exposure. The effects of these psoralens when administered orally
were also examined. The oral dose of each drug was 0.6 mg/kg,
administered 2 hr prior to laser exposures. The presence of these
psoralens markedly reduced the MED for delayed erythema, and the
maximum erythema response typically occurred 48-72 hr after exposure. Our results are entirely consistent with responses to these
psoralens plus conventional UVA sources. Reciprocity was also
observed to hold within experimental error for delayed erythema
of psoralen-sensitized skin. The nitrogen laser exposure doses
delivered at 106 mW/cm^2, necessary to induce immediate erythema,
delayed erythema with and without psoralens, IPD and tanning, are
summarized together in Figure 2.

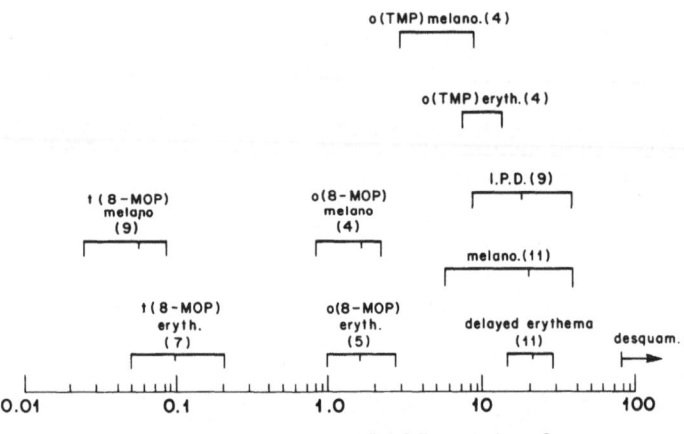

LASER EXPOSURE DOSE, J / cm²

Fig. 2. Laser exposure thresholds for cutaneous reactions in
fair-skinned Caucasians. Parentheses indicate number of subjects.
Brackets indicate range of values while the bars within the brackets indicate the average. Abbreviations: eryth = erythema; melano =
true melanogenesis; 8-MOP = 8-methoxypsoralen; TMP = trimethyl-
psoralen; o = oral administration of 40 mg to adult volunteers;
t = topical application of 0.1 ml per sq in. of 1% solution;
IPD = immediate pigment darkening; desquam = desquamation.

Histologically, the effects of the nitrogen laser at both 24
hr and 48 hr after exposure were similar to those of broadband
UVA from a conventional compact xenon arc source. Equally erythemogenic matched sites on the buttocks were chosen for comparison

of histologic damage as visualized by routine H&E staining and
light microscopy. At 24 hr after exposure, the vessels of the
dermis showed signs of vasodilitation and a perivascular lympho-
cytic infiltration was noted. At 48 hr, scattered dyskeratotic
keratinocytes were observed in addition to the above dermal changes.

Although we failed to find any significant deviations from
reciprocity or other unusual responses to this laser, it is apparent that the nitrogen laser is a very useful source for quantifying the effects of photosensitizing compounds active at 337.1 nm.
In addition to the psoralens studied above, phototoxic indices of
various crude coal tar preparations (5% crude coal tar in petro-
latum, Estar[R], and Zetar[R]) used in tar photochemotherapy of psoriasis were readily determined using the N_2 laser. The phototoxic
index at 337.1 nm was taken to be the ratio between the subject's
delayed erythema MED without tar to that with tar. These clinically used preparations all exhibited a phototoxic index between
2.5 and 4.0.

KrF Excimer Laser. Because of the high exposure doses necessary to induce cutaneous responses at wavelengths longer than
320 nm, it was not possible to study the effects of a single N_2
laser pulse of sufficient energy to cause observable responses.
However, the skin is much more sensitive to shorter UV wavelengths,
and hence the KrF excimer laser (248 nm) makes such a study possible. This laser (Tachisto, Inc. 150 XR) produces up to 100 mJ
per pulse with a pulse width of 30 ns. The peak irradiance was
approximately 10^7 W/cm^2. An initial study comparing the MED of
1 to 4 KrF laser pulses, with the MED of low-power, continuous
254 nm irradiation from a low-pressure mercury discharge lamp,
was conducted in 5 albino guinea pigs. The animals were depilated
by warm wax treatment 48 hr before exposure. A series of sites
12 mm^2 in area were exposed to a series of exposure doses, with
248 nm pulsed laser exposures on one side of the back and 254 nm
continuous-source exposures to corresponding sites on the other
side of the back. Erythemal responses recorded at 8 hr after the
exposure varied markedly between animals, but within each animal
the MED to the laser and to the 254 nm source were comparable at
approximately 100 mJ/cm^2.

Thus far, we have performed an analogous experiment in only
one lightly-tanned human subject using adjacent sites of the inner
forearm. As has been previously reported, the human MED is generally lower than that of the guinea pig for this spectral region.
In the one human tested thus far, the MED to KrF laser was attained in a single 30 ns pulse of 34 mJ/cm^2, while the 254 nm MED was
25 mJ/cm^2, delivered over several minutes. These values are within experimental and radiometric error. The minor difference noted
may also reflect some decreased sensitivity at 248 nm compared
with that at 254 nm. Although this study is still in progress,
it appears that, over 10 orders of magnitude in irradiance, UVC-
induced delayed erythema obeys reciprocity.

ArF Excimer Laser. Finally, we have investigated delayed erythemal response to the ArF excimer laser at 193 nm (Tachisto, Inc.,
150 XR). Little is known about cutaneous responses to wavelengths
less than 230 nm. An exposure dose of 500 mJ/cm^2 in the human and

2.1 J/cm^2 in the guinea pig elicited no clinically observable response. These doses are 20 times the MED at 248 nm. The optical density of the stratum corneum at 193 nm is 15 to 20 times that at 248 nm, because of strong absorption by peptide bonds. The result that a high exposure dose of 193 nm laser irradiation does not elicit a response is therefore reasonable if one assumes that delayed erythema is a response following absorption of photons within the living cell layers of the skin, which are shielded from 193 nm radiation by absorption in the dead cell layer, the stratum corneum.

Summary

The results and experiments presented here are a summary of ongoing attempts to observe reciprocity failure in human skin and some uses of the laser as a tool for studying phototoxic drugs. We have observed that, for delayed erythema of human skin, reciprocity holds for peak irradiances up to 2×10^6 W/cm^2 at 337.1 nm and 10^7 W/cm^2 at 248 nm. Such work is essentially impossible without lasers. As greater peak power levels become available in UV lasers, it may be possible to observe and study reciprocity failure in human skin in greater detail. We have also tried to point out the usefulness of lasers in studying the effects of photosensitizing drugs or compounds.

References

1. Claesson S, Juhlin L, Wettermark G: The reciprocity law of UV radiation effects. Acta Dermato-Venereol 38:123-136, 1958.

2. А. Г. ГАВРИЛОВ, Т. Н. МЕНЬШОНКОВА. Н. Ф. ЦИСКУНКОВА, М. Е. ПОСПЕЛОВ, Г. Я. ФРАЙКИН, Л. Б. РУБИН : О СПЕЦИФИКЕ ДЕЙСТВИЯ ЛАЗЕРНОГО У.-Ф. ИЗЛУЧЕНИЯ НА ВЫЖИВАЕМОСТЬ МИКРООРГАНИЗМОВ . Доклады Академии наук СССР 1978. Том 239, № 5

3. Johnston B, Johnson MA, Moore CB, Hearst JE: Psoralen-DNA photoreaction: Controlled production of mono- and diadducts with nanosecond ultraviolet laser pulses. Science 197:906-908, 1977

4. Grossweiner LI, Smith KC: The role of DNA polymerase I in photosensitization of E. Coli K-12 by 8-methoxypsoralen. Program and Abstracts, 7th Annual Meeting of the American Society for Photobiology, Asilomar, Calif., June 1979, p 140

5. Bucker H: Zur Abgrenzung des UV-Erythems durch das unspezifische Strahlungserythem. Strahlentherapie 777:404-413, 1960

6. Pathak MA, Hori Y, Szabo G, Fitzpatrick TB: The photobiology of melanin pigmentation in human skin, in Biology of Normal and Abnormal Melanocytes (T Kawamura, TB Fitzpatrick, M Seiji, editors). University of Tokyo Press, Tokyo, 1971, pp 149-167

Helium-Cadmium Laser (325 nm) Irradiation of Pigmented and Unpigmented Mammalian Skin: Electron Microscopic Effects on Dermal Nerve Fibers

R.I. Garcia and G. Szabo

Laboratory of Electron Microscopy, Harvard School of Dental Medicine
Boston, MA 02115, USA

1. Introduction

The application of lasers to dermatologic research and clinical practice
has been the subject of intensive investigation and has clearly been shown
to have great promise [1, 2]. Particularly, the pioneering work of GOLDMAN
[1-7] has convincingly demonstrated the effectiveness of the clinical use
of lasers in dermatology.

Almost all investigations to date on skin have used visible light or
infrared lasers; most commonly the pulsed ruby laser with output at 694 nm.
It has been shown that following ruby laser irradiation there is damage to
the epidermis and to the superficial connective tissue, which is followed
by healing of the 'wound' site [3, 8-10]. Of particular interest to us is
the finding that heavily pigmented skin is selectively damaged by the ruby
laser output [4]. This greater susceptibility of colored structures to
ruby laser damage has been effectively exploited [1, 2, 6, 11] . An example of this is the successful application of ruby laser in the treatment of
port-wine lesions [7]. However, it is evident that this selective damage
of pigmented tissues may often be a thing that one needs to avoid, particularly in using lasers on patients with normally darkly pigmented skin.

Melanin pigmentation affords natural protection to the skin against
harmful effects of solar ultraviolet radiation. The protective actions
of melanin in skin and the role of melanin in cellular physiology have been
the subjects of a number of reviews [12, 13]. We are interested in what
the effects of ultraviolet laser radiation would be on skin, particularly
in comparing the responses of pigmented and unpigmented skin to UV laser
irradiation. Would the protection melanin affords against natural UV be
duplicated in response to laser UV? Further, we are interested in finding
to what degree could these various laser effects on skin be attributed to
wavelength effects. In this regard, ROUNDS [14] has found that, based on
cell culture experiments and using morphologic and biochemical data, the
cytotoxic effect produced by laser radiation appears to be a wavelength-
specific phenomenon.

SZABO ET AL. [15] have compared the effect of 325 nm UV laser radiation
on pigmented and unpigmented guinea pig skin and have shown that melanin
appears to selectively protect pigmented epidermal cells and the superficial dermis from UV laser radiation, an opposite result from that obtained
with the use of ruby laser. We have, in this report, expanded on our earlier investigation to examine possible selective effects of UV laser on
dermal nerve fibers.

2. Materials and Methods

Guinea pig ear skin was irradiated with 325 nm ultraviolet, using a Spectra Physics helium-cadmium laser, at 15 mW with a beam diameter of 1.5 mm, for 16 to 30 minutes. A total of seventeen guinea pigs were used: white, tan, or black in color. Biopsies were obtained from both the irradiated site and an adjacent unirradiated site from each animal: immediately after radiation, 24 hours, 72 hours, and 30 days after radiation. Sodium bromide split skin preparations were treated with dihydroxyphenylalanine to visualize melanocytes and studied with the light microscope as whole mount preparations and another portion of each biopsy was fixed for electron microscopy and embedded in Epon. The embedded tissue was sectioned with a Porter-Blum MT-2 ultramicrotome and stained with lead citrate and uranyl acetate, and examined in AEI 6B and AEI Corinth 275 electron microscopes.

3. Results and Discussion

Damage produced by UV laser irradiation was easily recognizable at the light microscopic level in sodium bromide split epidermis and in conventional histologic sections, and also easily discernable at the electron microscopic level. In albino and lightly pigmented skin, immediately after irradiation, there is vacuolization of epidermal cells and development of small edematous lesions in the papillary dermis. Small focal dermal areas, believed to be occupied by peripheral nerves, were also injured and the myelin sheaths of larger dermal nerves also showed evidence of damage. By 24 hours post-radiation, an extensive dermal edema develops and there is often an infrabasal separation of the epidermis from the dermis followed by necrosis of the overlying epithelium. At the periphery of lesions there is varying degree of epidermal damage, including cellular vacuolization and destruction of nuclei. The lesion 24 hours post-radiation is approximately 1 mm in diameter in both the epidermis and the dermis, corresponding to the beam diameter. Besides the damage to neural elements found in the white and lightly pigmented skin, there is also a replacement of the collagen bundles in the dermis by a homogeneous, eosinophilic material.

An important feature of the lesion produced by the UV laser is the very circumscribed area of damage, which corresponds to the beam diameter. This may well be due the 'non-thermal' nature of the UV laser effects. As an elevation of less than 2 ºC was measured during irradiation, the effects we have observed are presumed to be primarily photochemical in nature and not photothermal.

Fig.1 Electron micrograph of black guinea pig ear skin UV laser irradiated. The epidermal basal layer and the superficial dermis show no evidence of laser damage. The subepidermal unmyelinated nerve is intact and has a normal appearance. K=keratinocyte, N=Schwann cell enveloping nerve axons. Magnification= X 4500.

Fig.2 Electron micrograph of black guinea pig ear skin UV laser irradiated. Intact dermal nerve with a normal appearance showing no evidence of damage from UV laser, illustrating the effect of melanin in shielding dermal structures from radiation. Here both myelinated (M) and unmyelinated (U) nerve axons are seen together. The appearance of the nerve fibers illustrated in Figs. 1 and 2 is markedly different from the nerve fibers in Figs. 4 and 5, which are taken from UV laser irradiated white guinea pig ear skin and which show marked evidence of cell damage. Magnification= X 12000.

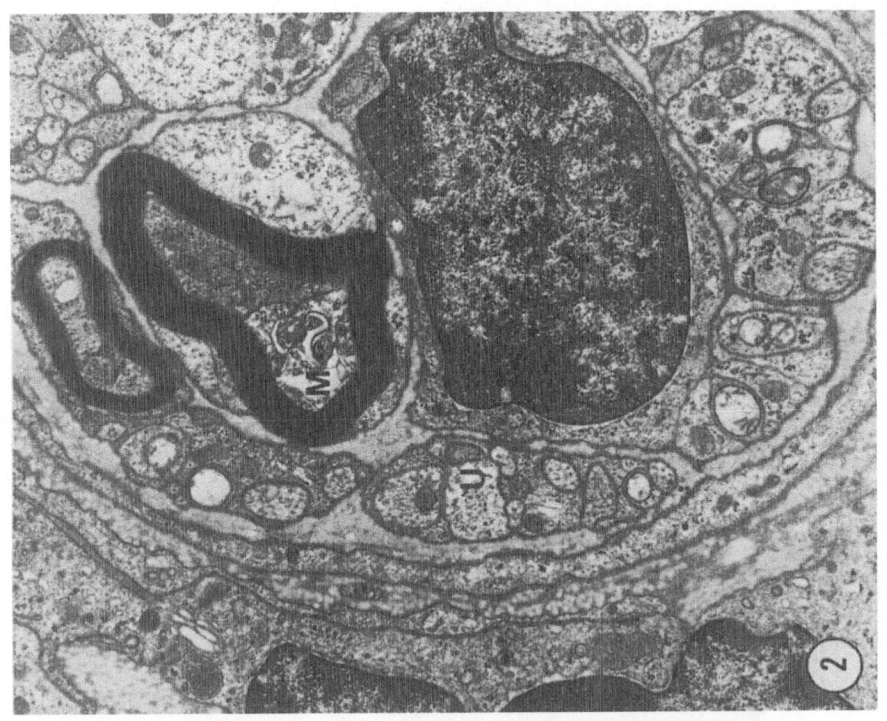

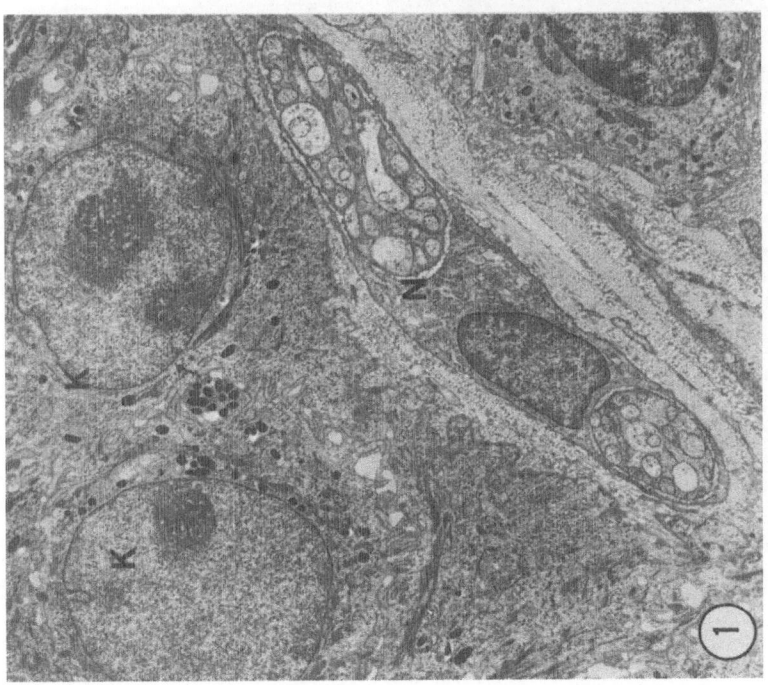

Figs. 1 and 2. Figure captions see opposite page

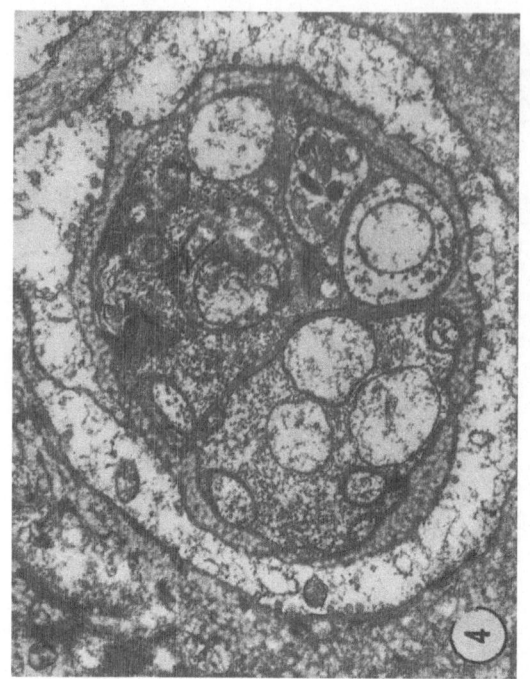

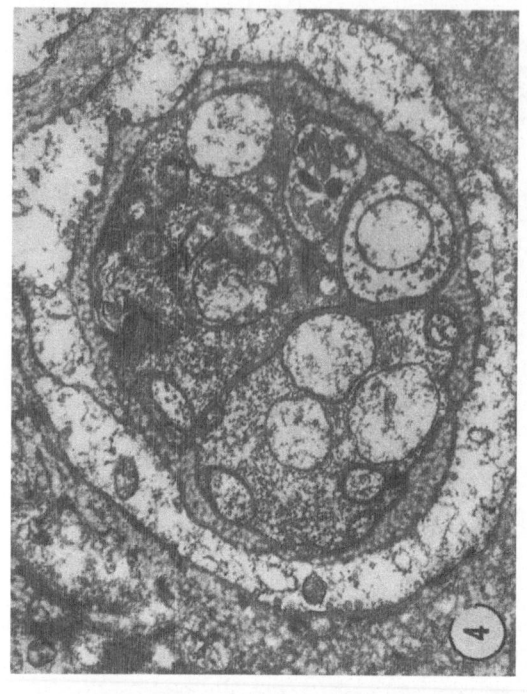

Fig.4 (on previous page) White guinea pig ear skin UV laser irradiated. The unmyelinated nerve seen and the surrounding dermis show signs of laser damage. Magnification= X 16000.

Fig.3 (on previous page) White guinea pig ear skin, unirradiated control. The unmyelinated nerve seen is normal and may be compared to the UV laser damaged nerve in Fig. 4. Magnification= X 12000.

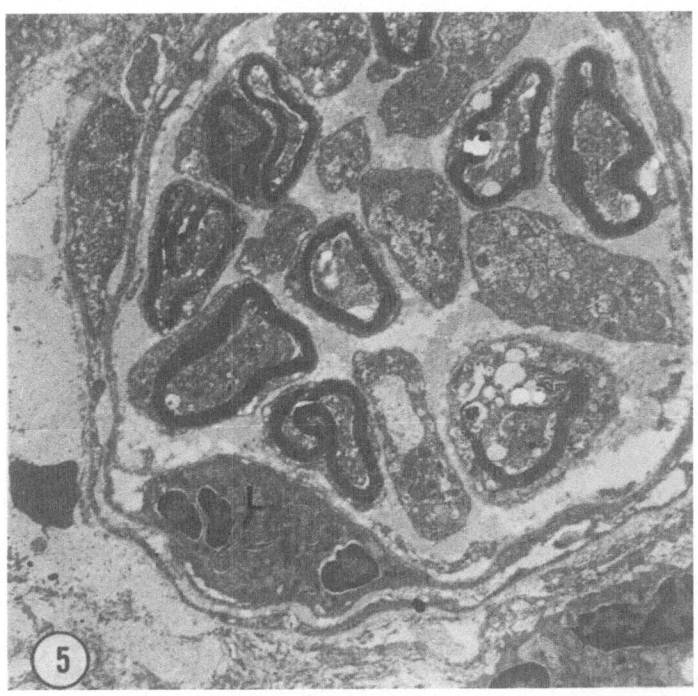

Fig.5 (above) White guinea pig ear skin UV laser irradiated. This larger
nerve has both myelinated and unmyelinated axons. The UV laser radiation
results in a disruption of the dermal collagen bundles, vacuolization of
Schwann cell cytoplasm, alteration of axon membranes, and a disruption of
myelin sheaths of nerve fibers. L=leucocyte. Magnification= X 4500.

In contrast to the results found in white and lightly pigmented skin,
the high melanin content of darkly pigmented skin was found to afford protection from UV laser radiation [15]. In particular, we have found that
there is much less damage to the connective tissue in the darkly pigmented
skin, possibly due to the 'shielding' effect of the epidermal melanin. It
is noteworthy that the myelin sheaths of dermal nerve fibers, and also the
unmyelinated nerve endings, appear to be selectively damaged by UV laser.
This is particularly evident in the cases where the melanin content of the
epidermis is not sufficient to block the UV radiation from penetrating to
the dermis. The susceptibility of the dermal nerve fibers appears to be
greater than that of other cellular constituents of the dermis to UV laser.

The damage to the dermal nerve fibers resulting from UV laser radiation
may be seen in Figs. 4 and 5. In comparison to the normal appearance of
nerve fibers in Figs. 1-3, the damaged nerve fibers are characterized by
membrane alterations, cytoplasmic vacuolization, and particularly striking
in the myelinated fibers is the disruption of the myelin sheath. These
effects of the UV laser are not found in those cases where the epidermal
melanin content is sufficient to screen out the radiation from penetrating
to the dermis. It is still unclear to what degree the dermal nerve fibers
are more susceptible to UV laser radiation than other dermal cells. The

literature on laser effects on neural cells and tissues is concentrated on visible and infrared lasers and relatively little work has been done with ultraviolet lasers, particularly at 325 nm. There are a number of reports dealing with ruby laser damage to nerves and the special sensitivity of cellular membranous structures [2, 6, 11].

The further application of UV lasers to skin research will prove valuable in answering many questions of interest to photobiology of the skin, especially in elucidating the effect of natural solar ultraviolet on skin. It appears from our experiments [15] that the helium-cadmium laser exerts a comparatively mild and localized effect on skin. Its use in dermatologic surgery may be recommended, as the healing process is rapid and there is no selective damage to melanized cells.

4. Acknowledgements

We thank our collaborators, S. Fine and D. MacKeen, who also served as co-authors on our earlier publication on this same subject, and we gratefully acknowledge the assistance of E. Flynn in the preparation of the electron micrographs.

Supported by USPHS Grant AM 20669 from the National Institute of Arthritis, Metabolism, and Digestive Diseases.

5. References

[1] Goldman, L., N.Y. State J. Medicine (1977) 77:1897-1900

[2] Goldman, L., and R. J. Rockwell, Jr., Lasers in Medicine, Gordon and Breach, N.Y., 1971, pp.385

[3] Goldman, L., D. J. Blaney, D. J. Kindel, D. Richfield, and E. K. Franke, Nature (1963) 197:912-914

[4] Goldman, L., Fed. Proc. (1965) 24:(Suppl. 14) S92-S93

[5] Goldman, L., (Ed.) Second Conference on the Laser, An.N.Y.Acad.Sci. 163: 383-720 (1970)

[6] Goldman, L., Biomedical Aspects of the Laser, Springer-Verlag, Berlin, 1967, pp.232

[7] Goldman, L., Arch. Dermatol. (1977) 113:504-505

[8] Fine, S., E Klein, S. Farber, R. E. Scott, A. Roy, and R. E. Seed, J. Invest. Dermatol. (1963) 40:123-124

[9] Klein, E., S. Fine, Y. Laor, M. S. Litwin, J. Donoghue, and L. Simpson, (1964) J. Invest. Dermatol. 43:565-570

[10] Helwig, E. B., W. A. Jones, J. R. Hayes, and E. H. Zeitler, Fed. Proc. (1965) 24:(Suppl. 14) S83-S91

[11] Gamaleya, N. F., in: Laser Applications in Medicine and Biology (Ed., M. L. Wolbarsht) Plenum, N. Y., 1977, pp.1-174

[12] Fitzpatrick, T. B. (Ed.) Sunlight and Man, Univ. of Tokyo Press, Tokyo, 1974, pp. 870

[13] Garcia, R. I., G. Szabo, and T. B. Fitzpatrick, in: The Retinal Pigment Epithelium (Eds., K. Zinn and M. Marmor) Harvard University Press, Cambridge, Mass., 1979, pp.124-147

[14] Rounds, D. E., Fed. Proc. (1965) 24:(Suppl. 14) S116-S121

[15] Szabo, G., D. MacKeen, and S. Fine, Annal. Ital. Dermatol. Clin. Sper. (1974) 28:203-224

The Application of Dye Lasers on Photodermatology

A. Anders

Inst. für Biophysik, Universität Hannover, D-3000 Hannover, Fed. Rep of Germany

P. Aufmuth

Inst. A für Experimentalphysik, Universität Hannover,
D- 3000 Hannover, Fed. Rep. of Germany

E.-M. Böttger and H. Tronnier

Hautklinik der Städtischen Kliniken, D-4600 Dortmund 1, Fed. Rep. of Germany

1. Introduction

In this review, the use of dye lasers as light sources in photodermatology is discussed. Besides the high spectral intensity
and narrow bandwidth, dye lasers offer the additional advantage
of tunability. We report on investigations on the erythema and
the treatment of psoriasis with and without the photosensitizer
8-methoxypsoralen (8-MOP).

In the literature the erythema effectiveness curve is a subject of controversy [1]. Two features are responsible for this:
the erythema shows a strong dependence on the bandwidth of the
irradiation source [2] and on the time of measurement [3]. These
problems have been cleared up by irradiating with narrowband
dye lasers and comparing results obtained at the same time after
irradiation. The erythema effectiveness curve with and without
8-MOP, the minimal erythema dose, and the erythema gradation
were investigated.

The advantage of narrow-bandwidth irradiation in phototherapy
and photochemotherapy has been tested with the aim of reducing
possible side effects and the total intensity while exciting at
the maximum of the action spectrum. Therefore, in previous investigations absorption spectra and transmission changes of human skin with and without photosensitizers were studied as the
basis for a wavelength-selective therapy of this kind [4-6].

2. Experiments

We applied a flashlamp-pumped pulsed dye laser (Chromatix CMX-4)
with a spectral bandwidth of about 0.06 nm in the UV region. The
energy and power per pulse amounted to 0.4 mJ and 0.4 kW respectively at 300 nm, i.e. the pulse length was 1 μs. The repetition
rate was 25 pulses per s. The laser beam was directed on to the
patients; the irradiation areas varied from 1 to 3 cm diameter.

The erythema reaction was produced by irradiation in the wavelength range from 260 to 345 nm. The degree of the erythema was
subjectively determined in mainly six gradations after 24 hours,
the uncertainty being one half of these steps. The irradiation

of psoriatic skin was performed at distinct wavelengths and repeated about every day over a period of one to two weeks. The
recovery process was controlled measuring the oxygen partial
pressure $p(O_2)$ in the skin [7,8].

Further details of the experimental conditions are described
by ANDERS et al. [2].

3. Erythema Investigations

The erythema measurements were performed without a photosensitizer (four normal persons with different sensitivity of the
skin) as well as after oral administration of 8-MOP (two persons). The latter ones received 60 mg 8-MOP 2 to 3 hours before
irradiation, as customary in phototherapy. The irradiation areas
were situated on the back of the patients.

Between 290 and 302 nm the erythema reaction was measured in
steps of 1 nm, outside this region every 5 nm. Figure 1 shows
the erythema effectiveness curve for untreated skin for two
different irradiation intensities, Fig. 2 after treatment with
8-MOP. A relatively sharp maximum appears around 298 nm, slightly shifted and with a higher peak value with 8-MOP. The minimal
erythema dose (MED) with and without 8-MOP is demonstrated in
Fig. 3.

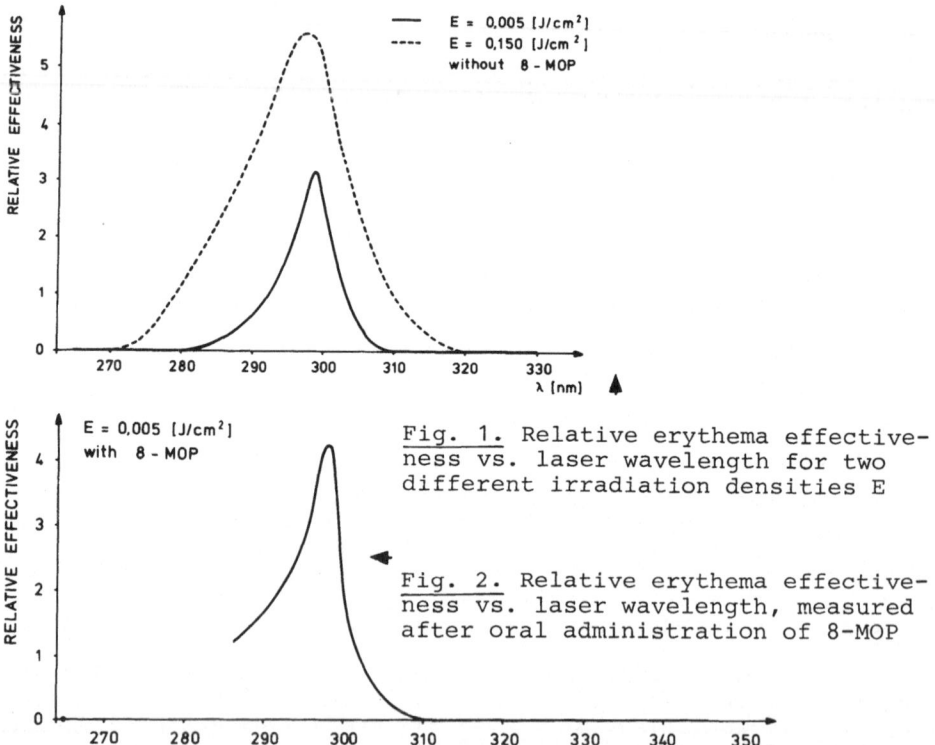

Fig. 1. Relative erythema effectiveness vs. laser wavelength for two
different irradiation densities E

Fig. 2. Relative erythema effectiveness vs. laser wavelength, measured
after oral administration of 8-MOP

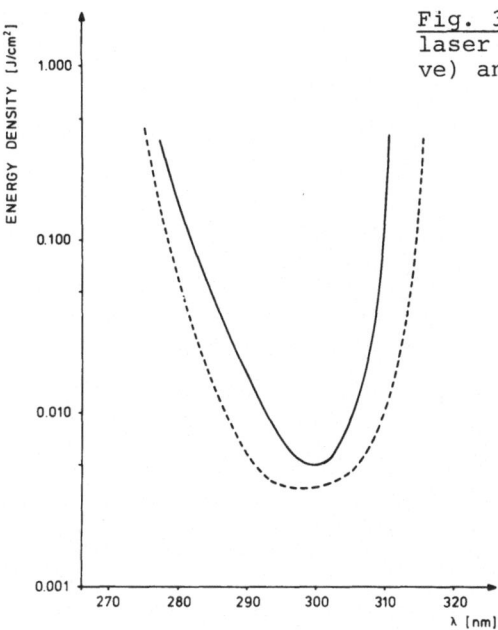

Fig. 3. Minimal erythema dose vs. laser wavelength without (solid curve) and with (dashed curve) 8-MOP

Between 260 and 280 nm, there is no erythema reaction with an irradiation intensity of 0.005 J/cm^2 corresponding to the MED at 300 nm; even with an intensity of 0.15 J/cm^2 at 270 nm no erythema was found. Up to now, erythema reactions for wavelengths shorter than 260 nm could not be observed because of the limited range of the laser used.

After application of 8-MOP no pigmentation was found between 310 and 345 nm, even with an intensity of 0.5 J/cm^2. This is surprising because at 365 nm there is supposed to be a maximum of the action spectrum of photosensitization with 8-MOP [9], but our result is similar to the action spectrum of the photo-reaction between 8-MOP and DNA in vitro which shows a maximum near 310 nm and a decrease towards longer wavelengths [10]. On the other hand, the quantum yield of this reaction increases from 260 to 370 nm [10]. PARRISH et al. [11] found a MED of 0.05 to 0.2 J/cm^2 at 337 nm after application of 8-MOP using an N$_2$ laser.

In Table 1 our MED values are compared to those from the literature. The MED seems to be reduced when a narrowband irradiation source is used; a similar tendency towards lower erythema thresholds was observed by SCHALLA et al. [12].

The dependence of the erythema reaction on the irradiation intensity, the gradation curve, is demonstrated in Fig. 4 showing the same behavior as with broadband irradiation [3].

Table 1 Minimal erythema dose (MED) and 50 % bandwidth of irradiation

MED $[mJ/cm^2]$	λ [nm]	$\Delta\lambda$ [nm]	Author
1 - 5	297	0.06	ANDERS et al. [2]
6 - 24	297	2.2	BERGER et al. [3]
10	300	5	FREEMAN et al. [14]
14 - 34	297	10	EVERETT et al. [15]

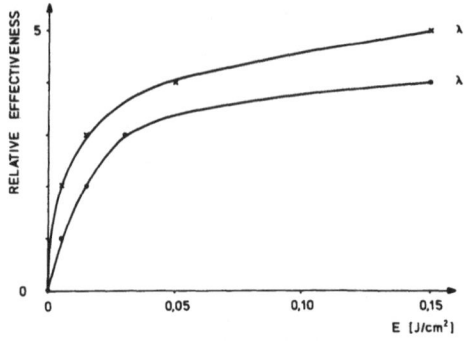

λ = 300 nm

λ = 292 nm

Fig. 4. Erythema gradation; relative erythema effectiveness vs. irradiation intensity for two wavelengths

4. Discussion of the Erythema Effectiveness Curve

Figure 5 gives a comparison of our results with those obtained by other authors [2,3,13-15]. The erythema curve with a peak near 300 nm, a minimum at 280 nm and a second but lesser maximum at 250 nm was accepted as a standard in 1935, but newer results have given different curves without a minimum at 280 nm or with only a slight minimum. Our results, however, reproduced nearly the standard curve until about 280 nm.

Such controversial measurements can be explained in two ways. The form of the curve of EVERETT et al. (see Fig. 5) originates from being measured 7 hours after irradiation, whereas the other curves were obtained 24 hours after irradiation. The differences can be explained by the temporal development of the erythema, which was studied by BERGER et al. [3].

The differences between those curves measured at the same time after irradiation, arise from the marked dependence of the erythema production on the bandwidth of the radiation source. This effect is represented in Fig. 5 in the depth of the minimum between 250 and 300 nm. The smaller the bandwidth of the source the more pronounced the minimum appears. Theoretical considerations about the variation of erythema with the bandwidth of irradiation were published by DIFFEY [16] and VAN DER LEUN [17].

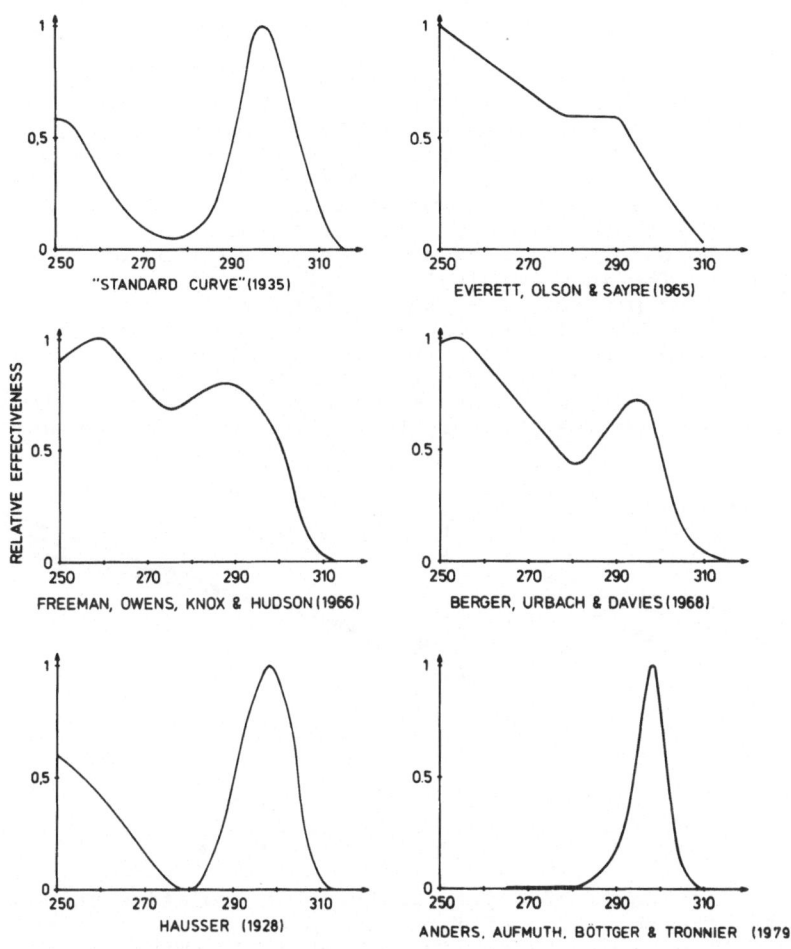

Fig. 5 Comparison of relative erythema effectiveness curves of various authors. All observations after 24 h, except EVERETT et al. (7 h). Irradiation bandwidth: HAUSSER [13]: 0.2 nm; EVERETT et al. [15]: 10 nm; FREEMAN et al. [14]: 5 nm; BERGER et al. [3]: 2.2 nm; ANDERS et al. [2]: 0.06 nm

With broadband irradiation the erythema production is influenced predominantly by the adjacent maximum of the curve, e.g., at 290 nm the broadness of the intensity distribution leads to a strong erythema production caused by the 300 nm maximum, because of the greater sensitivity of the skin in this region. To avoid such overlapping effects, a laser seems appropriate because of its narrow bandwidth. - Fifty years ago, HAUSSER obtained almost our erythema curve because he irradiated with distinct mercury spectral lines whereas the other curves were obtained with broad xenon lamps.

5. Treatment of Psoriasis

We irradiated five patients, two of them after oral application of 8-MOP (60 mg per person, 2 to 3 hours before irradiation; see above). Figure 6 shows the dependence of the oxygen pressure in psoriatic lesion after treatment with three different laser wavelengths, the irradiation intensities are given in the figure caption. The measured values were normalized to the mean value for healthy skin (80 Torr corresponding to 1), and a straight line was fitted to the points assuming a standard deviation of 20 % for each of them. The increase in the $p(O_2)$ values seems to represent the healing process. The healing without photosensitizer shows a correlation with the erythema curve; the best result was observed with an irradiation wavelength of 300 nm. After application of 8-MOP the effective wavelength region is extended. Further measurements are necessary to gain a deeper insight into the relationship between the erythema curve and the psoriasis healing process.

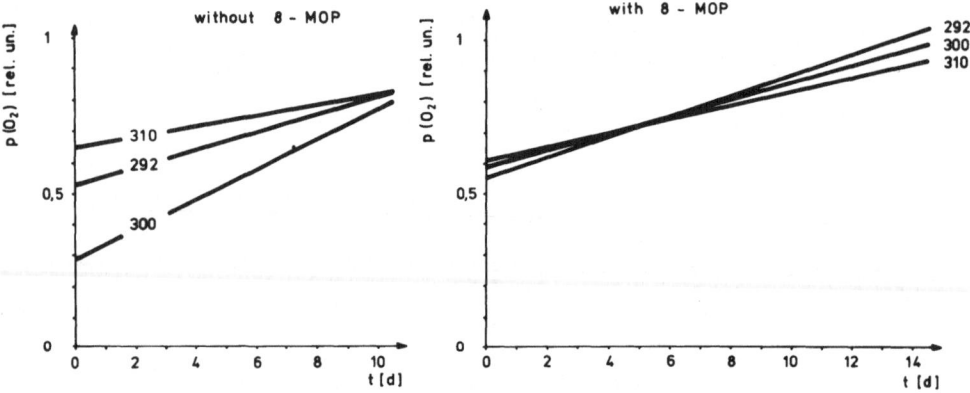

Fig. 6 Partial pressure of oxygen in the skin (relative values) vs. time after the beginning of irradiation for three different wavelengths without and with application of 8-MOP. Typical irradiation intensities per day: $0.01 \ J/cm^2$ (292, 300 nm), $0.15 \ J/cm^2$ (310 nm) without 8-MOP; $0.01 \ J/cm^2$ (292, 300 nm), $0.03 \ J/cm^2$ (300 nm) with 8-MOP

6. Concluding Remarks

The action spectrum governing the healing process of psoriasis is of great practical interest. A particular region lies around 300 nm because of the erythema maximum and the nucleic acids' absorption in this spectral range (Fig. 7, upper part). At the same time this is the region where mutations in DNA occur under natural conditions from sunlight (Fig. 7, solar spectrum).

In the lower part of Fig. 7 narrowband and broadband irradiation is compared to the broadness of the erythema curve. Because of the measured correlation between the erythema production and the healing process (without 8-MOP), the relatively narrow wave-

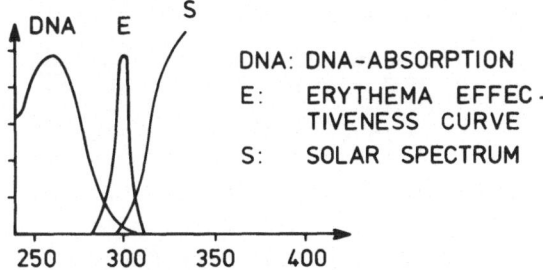

DNA: DNA-ABSORPTION

E: ERYTHEMA EFFEC-
TIVENESS CURVE

S: SOLAR SPECTRUM

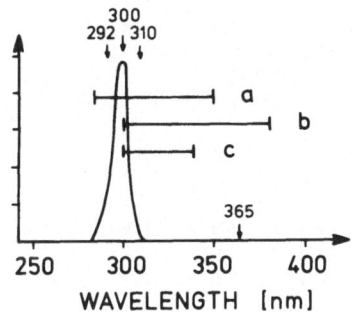

WAVELENGTH [nm]

Fig. 7. Upper part: DNA absorption, erythema effectiveness curve,
and solar spectrum. Lower part:
erythema effectiveness curve;
arrows mark the three irradiation
wavelengths used and the maximum
of the action spectrum of 8-MOP
photosensitization (see text); the
bars (a, b, c) indicate the spectral
ranges of usual lamps for psoriasis
treatment

length range of the optimum of healing suggests a narrowband irradiation in the phototherapy of psoriasis. The lamps normally
used for psoriasis treatment, however, have minimal spectral
ranges from 40 to 80 nm (see Fig. 7).

A further point to be considered is the degree of the penetration of light into the skin which decreases towards shorter
wavelengths (cf. Fig. 1 in ANDERS et al. [18].

Acknowledgement

We gratefully acknowledge Prof. Dr. A. Steudel for his interest
in our work and for discussing the manuscript.

References

1. J.H. Epstein: In The Science of Photobiology, ed. by K.C.
   Smith (Plenum Press, New York 1977) pp. 175-208
2. A. Anders, P. Aufmuth, E.-M. Böttger, H. Tronnier: In Laser
   79 - Opto-Electronics, ed. by W. Waidelich (IPC Science and
   Technology Press, Guildford 1979) pp. 355-360
3. D. Berger, F. Urbach, R.E. Davies: In XIII. Congressus In-
   ternationalis Dermatologiae, ed. by W. Jadassohn and C.G.
   Schirren (Springer, Berlin 1968) pp. 1112-1117
4. A. Anders, H. Zacharias, P. Aufmuth: In Laser 77 - Opto-
   Electronics, ed. by W. Waidelich (IPC Science and Technology
   Press, Guildford 1977) pp. 520-526

5. A. Anders, I. Lamprecht, H. Schaefer, H. Zacharias: Arch. Derm. Res. 255, 211-214 (1976)
6. A. Anders, H. Zacharias, I. Lamprecht, H. Schaefer: In Medizinische Physik, Vol. 2, ed. by W.J. Lorenz (Hüthig Verlag, Heidelberg 1977) pp. 539-547
7. E. Jacobsen: Scand. J. Clin. Lab. Invest. 37 Suppl. 146, 35-36 (1977)
8. H. Tronnier, E.-M. Böttger, E. Hoffmann: Z. Hautkr. 54, 546-550 (1979)
9. J.D. Spikes: In Ref. [1], pp. 87-112
10. F. Dall'Aqua, S. Marciani, G. Rodighiero: Z. Naturforschg. 216, 667-671 (1969)
11. J.A. Parrish, R.R. Anderson, C.Y. Ying, M.A. Pathak: J. Invest. Derm. 67, 603-608 (1976)
12. W. Schalla, B. Schaarschmidt, A. Anders, H. Zacharias, H. Schaefer: Arch. Derm. Res. 261, 89 (1978)
13. K.W. Hausser: Strahlentherapie 28, 25-44 (1928)
14. R.G. Freeman, D.W. Owens, J.M. Knox, H.T. Hudson: J. Invest. Derm. 47, 586-592 (1966)
15. M.A. Everett, R.L. Olson, R.M. Sayre: Arch. Derm. 92, 713-719 (1965)
16. B.L. Diffey: Arch. Derm. 111, 1070-1071 (1975)
17. J.C. van der Leun: In Research Progess in Organic, Biological and Medicinal Chemistry, Vol. 3, part II, ed. by W. Gallo and L. Santamaria (North-Holland Publ. Co., Amsterdam 1972) pp. 711-736
18. A. Anders, P. Aufmuth, E.-M. Böttger, H. Tronnier: In this book

Photorecovery in Human Skin

H. van Weelden

State University of Utrecht, Institute of Dermatology, Catharijnesingel 101
Utrecht, The Netherlands

Summary

The phenomenon of photorecovery was investigated in human
skin. The effect of irradiation of human skin with shortwave
ultraviolet radiation is reduced by subsequent or simultaneous
exposure to UV-A. This could be demonstrated by exposing the
skin to the full light from fluorescent sunlamps (UV-B) followed
by or simultaneously exposed to the full light from fluorescent
UV-A lamps. Photorecovery could also be demonstrated by exposing
the skin to the radiation from germicidal lamps (UV-C) supple-
mented by fluorescent UV-A lamps.
Exposing the skin to UV-A prior to UV-C shows a normal addition
of the erythematogenic effects of UV-A and UV-C. The exposure
of the skin to UV-A prior to UV-B suggests however that there
is a slight reinforced addition or augmentation.
The phenomenon of photorecovery could not only be demonstrated
in the case of UV-erythema, but also in UV-B induced tolerance
of the skin to light; the development of such tolerance, too,
was reduced if the exposure to UV-B was followed by UV-A.

Introduction

Photorecovery of the erythemal effect of ultraviolet radiation
in human skin was reported by van der Leun and Stoop [1],
but is still under discussion. Other investigators [2,3,4] could
not reproduce these results and found no photorecovery, or even
the reverse. Recently we were able to confirm in a rather simple
way that there is photorecovery in human skin.
Whether or not there is photorecovery in human skin is interesting from a general biological point of view, and specifically
in photodermatology, because photorecovery might be an indica-
tion that a grossly observable reaction of the skin could be related to damage and repair of DNA. Photorecovery in human skin
shows a striking parallelism with photoreactivation in bacteria
and yeast. If and in what way UV-erythema and, more likely, light-
induced tolerance to light are related to DNA-damage, is still
unknown.

Materials and Methods

The volunteers, ranging in age from 20 to 37 years, were exposed to UV-radiation of three wavelength regions. Philips TL
colour 12 fluorescent tubes were used as UV-B sources; these

lamps emit mainly UV-B with a peak emission near 305 nm. The
UV-A source consisted of Philips TL colour 09 lamps with the
peak emission near 365 nm. Lowpressure mercury arcs of the type
Philips TUV were used as UV-C source; these lamps are frequently
used as germicidal lamps and emit mainly radiation of the wavelength 254 nm.
Each volunteer was given a series of test irradiations with UV-
radiation on the trunk or the forearm. In each series, nine areas
were exposed to UV-radiation with doses increasing in a geometri-
cal progression, with $\sqrt{2}$ as the ratio. The dose necessary to
produce a just perceptible erythema was used as a measure for the
effectiveness of the radiation; the effectiveness was defined as
the reciprocal of the minimal erythema dose (MED).
Always two series of test irradiations were compared with respect
to effectiveness.
In the first series of test irradiations the skin was exposed
either to UV-C or UV-B. In the second series the same type of
radiation was used, but now in combination with UV-A. This was
done by exposing the skin to UV-A simultaneously with or immediately after the test irradiations; in control experiments, the
UV-A was given prior to the test irradiations. The effect of the
added UV-A was expressed as the ratio of the minimal erythema
doses for both series. As the ratio of the successive doses in
the test series was $\sqrt{2}$, the ratio of the two values of the MED
could be calculated as:

$$\frac{MED_B}{MED_{B+A}} = 2^{-\frac{n}{2}}$$

in which n is the difference in the number of erythemas observed
in the two test series, in the example given the series with only
UV-B and the series with UV-B followed by UV-A. For each combination of the wavelength regions investigated, a group of 8 to 10
volunteers were taken. The average of the ratios over such a
group is considered to be typical for that particular combination
of wavelength regions.
A second part of the investigation dealt with the induction of
tolerance to light in the skin by UV-B; the question was whether
or not this induction of tolerance was subject to photorecovery.
UV-B induced tolerance to light was investigated in 10 volunteers by exposing a large area on the trunk to an overdose as
high as 3 MED of UV-B. A second area was exposed to exactly the
same dose of UV-B, but now followed by an exposure to UV-A. A
third area was not exposed at all and served as a control. After
these initial exposures the areas were regularly tested with UV-B
to determine the change of the minimal erythema dose with time.
Used were the same type of test series as described above. The
protection factor PF achieved by such an initial exposure can
be calculated as:

$$PF = \frac{MED_{exp.}}{MED_{unexp.}}$$

The protection factor achieved was determined by regular assess-
ments for several weeks. The protection reaches a peak about 10
days after the initial exposure and disappears completely within two months. The effect of the subsequent exposure to

UV-A on the tolerance induced by UV-B can be calculated as the ratio of the protection factors in both exposed areas.

$$\frac{PF_{B+A}}{PF_B} = \text{the average of the PF's in one volunteer.}$$

A ratio smaller than 1 would point to photorecovery.

Results

In experiments where the test series of UV-B were followed by or simultaneously exposed to UV-A, the average result was:

$$\frac{MED_B}{MED_{B+A}} = 0.76 \pm 0.04$$

Preliminary experiments had shown no difference in the effect of UV-A on the UV-B exposure if the skin was exposed to UV-A simultaneously or after the UV-B exposure. The fact that the average is smaller than 1 means a reduction of the effective dose by the illumination with longer wavelengths.
The dose of UV-A given in this test series amounts to 0.13 MED. The addition of the erythematogenic effects of UV-A and UV-B will have a tendency opposite to photorecovery. If we compensate for the erythematogenic effect of UV-A by subtracting it from the effect of UV-B, we even get:

$$\frac{MED_B}{MED_{B+A}} = 0.66 \pm 0.05$$

To determine whether or not the addition of UV-A has any effect on the effectiveness of UV-B also the sign test [5] was used. Using this type of test the effect was found to be significant at a level $p=0.05$. The above determined ratio shows the average of the magnitude of the reduction with the standard error of the mean.
Similarly the erythemas elicited by the UV-C were also subjected to photorecovery.

$$\frac{MED_C}{MED_{C+A}} = 0.71 \pm 0.1$$

The dose of UV-A administered was 0.01 MED; this is still about 10x more in energy than the UV-C administered. The compensated value is

$$\frac{MED_C}{MED_{C+A}} = 0.70 \pm 0.1$$

Also the effect of UV-A on the effect of UV-C was, using the sign test, significant at a level $p=0.05$

The reversed application of UV-A and UV-C showed no reduction at all and indicate that there is a normal addition of the ery- thematogenic effects of UV-A and UV-C.

$$\frac{MED_C}{MED_{A \to C}} = 1.11 \pm 0.1$$

The dose of UV-A administered is 0.15 MED and the value correc- tec for addition will be:

$$\frac{MED_C}{MED_{A \to C}} = 0.97 \pm 0.1$$

The sign test also showed no significant effect.
The exposure of the skin to UV-A prior to UV-B shows a slightly reinforced addition or augmentation.

$$\frac{MED_B}{MED_{A \to B}} = 1.29 \pm 0.06$$

The UV-B was preceded by 0.11 MED of UV-A. The value compensated for addition is:

$$\frac{MED_B}{MED_{A \to B}} = 1.16 \pm 0.06$$

The augmentation may be there, but is not as striking as repor- ted by other investigators. The sign test gave p=0.14, indica- ting that the augmentation effect was not significant at the level adopted for this investigation, p=0.05.
For UV-B induced tolerance to light there was a reduction of the development of tolerance when the exposure to UV-B was fol- lowed by an exposure to UV-A. The average over the 10 volunteers was:

$$\frac{PF_{B+A}}{PF_B} = 0.61 \pm 0.06$$

The sign test showed that the reduction was significant (p=0.05).

Discussion

It appears that both 254 nm erythema and 300 nm erythema are subject to photorecovery. Any significant difference between the magnitudes of the reduction of erythematogenic effects of UV-B and UV-C by a subsequent or simultaneous exposure to UV-A could not be found in this particular experiment.
Photoprotection by exposing the skin to UV-A prior to the UV-C or UV-B exposures did not occur. The amount of UV-A administered prior to the UV-C irradiation suggests that the addition is very

likely to be a normal addition. Preirradiation with UV-A prior to UV-B suggests a reinforced addition or augmentation as also suggested by van der Leun and Stoop [1], claimed by Willis et al. [2] and confirmed by Spiegel et al. [4]. The augmentation in our investigation is not even statistically significant, and certainly not as striking as claimed by some of the other investigators.

We were just lucky to find the phenomenon of photorecovery in human skin in such an easy way. Many of the experiments done by other investigators to establish if there is photorecovery were not successful because the light-source for the UV-B exposure probably contained already enough UV-A to cause photorecovery during the exposure. Subsequent exposure to an additional dose of UV-A showed nothing but normal addition of the erythemal effectiveness of both irradiations. Van der Leun and Stoop |1| have already pointed out that the amount of longwave ultraviolet and visible light in a high-pressure mercury arc was sufficient to cause photorecovery during the exposure. The relatively small amount of UV-A in the fluorescent UV-B tubes was apparently not enough to produce photorecovery.

Now that it has been confirmed that the phenomenon of photorecovery really does occur in human skin it is time to perform more thorough investigations. There is a need for a tunable monochromatic light-source to establish what wavelengths are capable to reactivate and what wavelengths can be reactivated. It is surprising that the effect of photorecovery is just as strong if the UV-A is given during the test series with UV-B as exposing the skin to UV-A afterwards. Probably it takes only a short time-interval to achieve photorecovery. To study the influence of the time-interval one should not only have a monochromatic light-source, but also monochromatic light of sufficient intensity.

Photorecovery could not only be demonstrated for UV-erythema, but also for UV-B induced tolerance of the skin to light: such tolerance, too, was reduced if the exposure to UV-B was followed by UV-A. Light-induced tolerance to light might be related to DNA-damage and possibly to carcinogenesis.

Photorecovery of the UV-B induced tolerance to light might point more directly to a possible partial repair of damaged DNA. Especially this last consideration; the possible relationship with DNA-damage and carcinogenicity makes it necessary that tunable monochromatic light-sources become available to skin photobiologists.

References

1. J.C. van der Leun, Th. Stoop: " Photorecovery of UV erythema", in *The Biologic Effects of Ultraviolet Radiation*, ed. by F. Urbach (Pergamon, Oxford 1969) p. 251
2. I. Willis, A. Kligman, J. Epstein: Effects of long ultraviolet rays on human skin: photoprotective or photoaugmentative? J. invest. Dermat. *59*, 416 (1973)
3. C.Y. Ying, J.A. Parrish, M.A. Pathak: Additive erythematogenic effects of middle-(280-320 nm) and long- (320-400 nm) wave ultraviolet light. J. invest. Dermat. *63*, 273 (1974)
4. H. Spiegel, G. Plewig, C. Hofman, O. Braun-Falco: Photoaugmentation. Arch. of Dermat. Res. *261*, 189 (1978)
5. W.H. Beyer: *Handbook of Tables for Probability and Statistics*, 2nd ed. (The Chemical Rubberco, 1966)

Part IV

Phototherapy of Hyperbilirubinemia

Introduction to the Mechanism of Phototherapy of Jaundice and Related Technology

T.P. Vogl *

Departments of Pediatrics and Radiology, Columbia University
New York, NY 10032, USA

Jaundice of the newborn (neonatal hyperbilirubinemia) has been recognized as a potentially serious condition of newborn infants since antiquity. More than 15 per cent of all newborns exhibit clinical (visually observable) jaundice during the first few days of life. Prior to the 1940s, when exchange transfusion was introduced as treatment, kernicterus (bilirubin encephalopathy) was a leading cause of neonatal death and cerebral palsy. Jaundice is a descriptive term for the yellow-orange coloration observed in the skin (and sclera of the eyes) of individuals who have accumulated the pigment bilirubin in their bodies from any of a large number of possible causes. In the newborn infant a complex series of factors including immaturity of the enzyme systems of the liver, short life span of fetal erythrocytes, birth trauma, lack of colonic flora, etc. all combine to raise serum bilirubin concentrations [1].

Bilirubin results from the catabolism of most porphyrins in the body; more than 80 per cent results from the action of the reticulo-endothelial system, principally the spleen, on senescent erythrocytes (red blood cells). The hemoglobin in these cells undergoes cleavage and the globin is separated from the heme. The heme moiety is split and the iron recovered. The resulting straight chain tetrapyrrole, biliverdin, is reduced to bilirubin which is bound to serum albumin and transported to the liver. There, in the healthy individual, it is conjugated with glucuronic acid and excreted into the bile as the water soluble diglucuronide. In the presence of an excess amount of bilirubin, a deficient amount of albumin, or the presence of other molecules that compete for the bilirubin binding sites on albumin, unbound bilirubin may be present in the serum. Unbound bilirubin can pass through the blood-brain barrier and cause neurological damage or death, possibly by interfering with oxydative phosphorylation.

It must be emphasized that the foregoing is a greatly oversimplified description of the process. Many other aspects including a history of hypoxia, acidosis, or hypothermia; some administered drugs; delayed passage of meconium; fetal-maternal blood type incompatibilities; etc. affect the clinical picture. All these may affect the rate of formation or excretion of bilirubin, or the competition for the albumin binding sites. For more details on the clinical and patho-physiological processes involved see references [1] through [7].

---

* Mailing address: 4857 Battery Lane, Bethesda, MD 20014, USA

The introduction of exchange transfusion, in which the vascular system is "washed out" with a volume of blood equal to twice or three times the blood volume of the infant, greatly reduced death and permanent morbidity due to bilirubin encephalopathy, although there is morbidity associated with the treatment itself. Consequently, when Cremer et al [8] published their findings that infants exposed to light, particularly blue light, exhibited markedly reduced jaundice, this form of therapy "phototherapy" rapidly gained acceptance. In large part, the reason for this rapid acceptance was the perception that light cannot be harmful to infants. The argument was that the entire phylogenic development of man occurred in visible light and since the doses applied were only one tenth the intensity of sunlight in temperate climates they were clearly benign.

By 1967 phototherapy was in use around the world. The literature was recommending that illumination levels in nurseries be raised so as to give "prophylactic" phototherapy and reporting that no serious side effects were noted -- the bandwagon was rolling.

In the material below, events since 1967 are reviewed with emphasis on the potential role of the physical scientists, particularly optical physicists, in phototherapy of the newborn and as examples of the problems that may be encountered in future photomedical research. Four areas are considered: 1) photobiochemical mechanism of phototherapy; 2) kinetics and physiology of bilirubin; 3) side effects of phototherapy; and 4) light sources, light measurement, and instrumentation.

1. Photobiochemistry

The primary question for the past 20 years has been: How does the light do it? The extensive effort of many researchers has gone into this question and a fair summary of twenty frustrating years of work is that bilirubin is a very reactive compound and that a broad range of reaction products can form in the presence of light depending on the solvent, the degree of oxygenation, and the presence of other molecules. Some examples of these efforts can be found in References [9] through [16]. Until 1978 the unexplained fact was that under phototherapy, both of jaundiced infants [17,18] and of Gunn rats [19] (a congenitally jaundiced strain of rat), a substance that appears to be unconjugated bilirubin is secreted into the bile in increased quantity within a few minutes of therapy initiation, whereas bilirubin concentrations in the serum decline only over a period of hours [20,21,22]. Finally late in 1978 the explanation became clear when McDonagh, Lightner, and Wooldridge [23] and Schmid [24] showed that the process of phototherapy is the rearrangement of internal hydrogen bonds in the bilirubin molecule. Under ordinary conditions bilirubin is so strongly internally hydrogen bonded (EE configuration) that it has been considered lipophyllic. The photons break the hydrogen bonds and render the bilirubin sufficiently polar (EZ configuration) to permit the liver to excrete it without conjugation. Suddenly everything is explained; but it took 20 years and had to await the development of high pressure liquid chromatography*. It was also not until 1979 that Broderson [25] questioned whether bilirubin is truely lipophilic and suggested that bilirubin damages brain cells by binding to phosphatidyl choline rather than lipids. This revised view of the mechanisms of kernicterus also explains the damage enhancing effect of prior acidosis.

2. The Kinetics and Physiology of Bilirubin

An aspect of phototherapy that involves the physical and mathematical sciences is the problem of the kinetics of bilirubin and the distribution of bilirubin among coupled body pools. An understanding of the relevant dynamics is sought so that a rational basis for the administration of phototherapy can be obtained. A well worn analogy is that if one wishes to minimize the effort involved in killing ants as they come crawling from under the kitchen door, one only needs to swat when an ant appears. In 1973 the National Academy of Sciences Committee on Photo-therapy in the Newborn [26] speculated that continuous phototherapy may not be optimal if one wishes to minimize the light dosage in order to minimize potential side effects (reviewed below). The argument goes as follows: The photochemical reaction (whatever it may be) is very rapid, probably on the order of nanoseconds. Yet the skin bleaches in about four hours and the serum bilirubin concentration drops yet more slowly [27,28]. If one therefore assumes that the rate limiting step is the diffusion, either of bilirubin or of photoproducts, to or from the site of the photochemical reaction, then light pulses timed to coincide with fully loaded bilirubin sites may be equally effective at a significantly lower dose, and therefore safer.

Vogl and coworkers [20,21,22,29,30] examined these questions, and the related questions of the dose-response curve, by exposing groups of Gunn rats to various light intensities and on/off cycles. Using blue as well as white fluorescent light sources and energy densities ranging from 1.4 - 0.09 mW/cm^2 in the 415 to 450 nm band an exponential decay in serum bilirubin levels was observed with a time constant of about 1.2 days. More than one time constant appeared to be present but was not detectable with statistical significance.

The problems of taking such experimental in vivo data and relating them to diffusion processes between body compartments has been recently reviewed [31,32, 33]. It is a far more difficult and challenging problem than appears at first glance due to two well known but often ignored problems: 1) The problem of obtaining slopes from noisy data and 2) the lack of orthogonality in the representation of a function by a series of exponentials.

Dose-response curves were also obtained from this data and the semilog-arithmic nature of the dose-response curve of serum bilirubin levels to light intensity demonstrated; results that were confirmed by Ballowitz et al [34]. Such problems may appear to be only of academic interest. However, their immediate applicability to patient care is discussed below.

3. The Side Effects of Phototherapy

When phototherapy was first introduced no side effects were expected for the reasons previously delineated and, possibly because none were expec-ted, none were found. As late as 1972, in a major review article [2] on neonatal hyperbilirubinemia six "hazards" of phototherapy were listed (see also [6, 35]):

1) possible damage to the eyes

2) transitory loose stools

3) transitory benign skin rashes

4) problems in determining the degree of jaundice clinically (visually) under phototherapy

5) the rare incidence of "Bronze Baby" syndrome [36], the accumulation in the circulation of a bronze colored pigment, still unidentified, that occurs primarily in the presence of cholestasis

6) electrical danger from improperly grounded equipment.

With respect to the then unknown products of photodecomposition, the author stated "No studies demonstrate that products of phototherapy are as dangerous as bilirubin". However, times and opinions were changing and in the same paper the author states that phototherapy should not be used prophylactically. Since then a number of side effects have been demonstrated:

1) Maurer has shown that phototherapy injured platelets [37,38].

2) Odell and Bratlid have shown that bilirubin interacts with erythrocyte membranes both in vitro and in vivo and damages the membrane, producing potassium leaks. [5,39].

3) Speck, Rosencranz, and Santella have shown that bilirubin induces photodegradation in DNA and that, in cell cultures, intermittent phototherapy damages DNA more than continuous phototherapy, presumably by interfering with the DNA repair mechanisms [40,41].

4) Aplin has shown that phototherapy reduces levels of immunoreactive Prostaglandin A [42].

5) Wu [43] and Bell et al [44] have shown that phototherapy, alone or with concomitant use of radiant warmers, markedly increases the insensible water loss of newborns, particularly prematures.

6) Speck has found (unpublished communication) that phototherapy lights affect IV nutrient solutions through the effect of the light on tryptophan and riboflavin.

7) Sisson has shown that phototherapy reduces riboflavin levels in newborns but warns agains replacement therapy during phototherapy.

8) The process of phototherapy reduces maternal-infant eye contact beacause the eyes are covered and thereby interferes with maternal-infant bonding.

Phototherapy was introduced 20 years ago. Consequently, the first cohort of infants who were exposed to this treatment have just entered the child bearing age when long term effects might be expressed. Another consideration is that the disturbances in the DNA, if they are manifest at all, may be expressed in the skin (the primary target of of phototherapy) by an increase in susceptibility to skin cancer. Since this is the most common form of cancer, a ten per cent increase in the incidence would probably not be noted, let alone related back 40 years or more to phototherapy during the neonatal period. These points are emphasized in order to call the attention of physicists to some of the problems inherent in biomedical research and its clinical applications.

4. Light Sources, Light Measurement, and Instrumentation

Unfortunately photometry is an unknown science to many clinicians and to an amazingly large number of engineers -- including many designers of biomedical instrumentation. This is illustrated by the fact that until Klein's plea "Shedding Light on the Use of Light" [45] alerted the pediatric community to the problem, phototherapy exposures were measured with photographic or other visual-spectrum-oriented light meters with no regard to the action spectrum of bilirubin (it was known to lie in the blue from Cremer's original work) or to the spectral response of the meter. Investigators then switched to spectroradiometers but were sometimes led astray by the manufacturers.

A particularly unfortunate example is the early, careful work on dose-response by Mims et al [46] that was rendered suspect by his published values of irradiance three orders of magnitude lower than was expected from their light sources. They used a commercial radiometer whose calibration was certified to be traceable to a National Bureau of Standards light source. What had happened? The instrument they had purchased consisted of a silicon photodetector with a series of filters for wave length separation. The blue filters leaked a little red; the silicon cell has far higher sensitivity in the red than in the blue; the NBS calibrated light source (tungsten iodide) radiates far more energy in the red than in the blue. Couple these effects and use them to measure a fluorescent lamp with a high blue/low red output and a three order of magnitude calibration error results. Shortly thereafter a commercial instrument came on the market specifically designed to measure phototherapy irradiance. It consisted of a detector and a relatively broad interference filter. The meter on the front of the instrument was calibrated in microwatts/cm^2/nm!

At present most researchers active in the field, both clinical and laboratory workers, are aware of these problems. However, other researchers who may just be entering the field of photomedicine may not yet be that knowledgeable.

The in vivo action spectrum of bilirubin was established in 1977, independently by Ballowitz [34] and Tan [47]. Ballowitz also presents an interesting composite curve of previously published action spectra. Today the use of the more accurate assay methods of McDonagh [19] together with a tunable laser light source would permit far greater spectral resolution of the action spectrum and of the dose-response curve than previously possible.

The history of the development of the phototherapy units presently in clinical use is instructive. These units usually consist of six to eight 20 watt fluorescent tubes spaced by a tube diameter and covered with a Plexiglass-G sheet (to absorb any ultraviolet radiation). How did this configuration emerge? Unfortunately not on the basis of any dose-response study but simply because this was the largest number of tubes that could easily be placed above an incubator without interfering with nursing care. It is important to remember that changing equipment or procedures, once introduced into clinical practice, is very difficult both because of economic constraints and because any change from the "usual and customary" practice renders it an experimental procedure requiring appropriate protection for human subjects including informed consent. Therefore it becomes increasingly important that equipment be optimized before its widespread introduction into clinical use.

Finally, the variation of dose-response with serum bilirubin concentration remains an open question. Ballowitz [34] presents data that suggests dose-response varies with serum bilirubin concentration in Gunn rats. The clinical study by Vogl et al [29] can be interpreted to suggest that at serum concentrations of 8 - 10 mg/dl (concentrations at which bilirubin barely pigments the skin) the effect of phototherapy is marginal. Research on this point is urgently needed since many low-birthweight newborns are being exposed to phototherapy at these serum bilirubin concentrations.

Conclusions

The therapeutic potential of light has fascinated people since antiquity and many beneficial effects of light have been claimed. Medical uses are presently being discovered in dermatology and cancer therapy and are entering the clinical armamentarium. It will reward the physical scientist entering the field to be particularly wary of bandwagon effects and to acknowledge that extrapolation and scaling are far riskier procedures in biomedicine than in optics. It is also important to constantly bear in mind that photons (and phonons) are drugs when they are used for therapy or diagnosis and that no therapy has ever been found (although many have made the claim) that does not have adverse side effects. Consequently a physicist, like every other member of a biomedical research team, becomes morally responsible for the Hippocratic imperative: Above all, do no harm.

BIBLIOGRAPHY

1. L. M. Gartner, L. Kwang-sun, S. Vaisman, D. Lane, and I. Zarafin, Development of Bilirubin Transport and Metabolism in the Newborn Rhesus Monkey. J. Pediatr. 90:513 (1977)

2. M. J. Maisels. Bilirubin. Ped. Clin. N. A. 19:447 (1972)

3. R. Schmid and A. F. McDonagh. The Enzymatic Formation of Bilirubin. Ann. N. Y. Acad. Sci., 244:533 (1975)

4. J. Jacobsen and R. Brodersen. The Effect of pH on Albumin-Bilirubin Binding Affinity. Birth Defects: Original Article Series 12:175 (1976)

5. G. B. Odell. Neonatal Jaundice, in Progress in Liver Diseases, V:457 (1976), H. Popper and F. Schaffer, Eds.

6. R. E. Behrman, A. K. Brown, M. R. Currie, L. C. Harber, J. W. Hastings G. B. Odell, R. Schaffer, R. B. Setlow, T. P. Vogl, and J. Wurtman. Preliminary Report of the Committee on Phototherapy in the Newborn. J. Pediatr. 84: 135 (1974)

7. A. K. Brown, W. W. Zuelzer, and A. R. Robinson. Studies in Hyper-bilirubinemia. II. Clearance of Bilirubin from Plasma and Extra-vascular Space in Newborn Infants During Exchange Transfusion. Am. J. Dis. Child. 93:274 (1957)

8. R. J. Cremer, P. W. Perryman, and D. H. Richards. Influence of Light on the Hyperbilirubinemia of Infants. Lancet 1:1094 (24 May 1958)

9. C. L. Kapoor, C. Murti, and P. Bajpai. Interaction of Bilirubin with Human Skin. N.E.J.M. 228:583 (1973)

10. R. Bonnett and J. C. M. Stewart. Photo-oxidation of Bilirubin in Hydroxylic Solvents. J. Chem. Soc., Perkins Trans. I:224 (1975)

11. C. L. Kapoor. Interaction of Bilirubin with Reconstituted Collagen Fibrils. Biochem. J. 147:199 (1975)

12. D. A. Lightner and G. B. Quinstad. Hematynic Acid and Propentdyopents from Bilirubin Photo-oxydation In Vitro. F.E.B.S. Letters 25:94 (1972)

13. D. A. Lightner. In Vitro Photooxydation Products of Bilirubin. In Phototherapy in the Newborn: An Overview, p.34. National Academy of Sciences, Washington, D.C. (1974)

14. J. D. Ostrow, C. S. Berry, and J. E. Zarembo. Studies on the Mechanism of Phototherapy in the Congenitally Jaundiced Rat. In Phototherapy in the Newborn: An Overview, p.74. National Academy of Sciences, Washington, D.C. (1974)

15. E. Gabagnati and P. Manitto. A New Class of Bilirubin Photoderivatives Obtained In Vitro and their Possible Formation in Jaundiced Infants. J. Pediatr. 83:109 (1973)

16. A. F. McDonagh. Phototherapy and Hyperbilirubinemia. The Lancet p339 (2 February, 1975)

17. H. T. Lund and J. Jacobsen. Influence of Phototherapy on the Biliary Excretion Patterns in Newborn Infants with Hyperbilirubinemia. J. Pediatr. 85:262 (1974)

18. J.D. Ostrow, C.S. Berry, R.G. Knodell, J.E. Zarembo. Effect of Phototherapy on Bilirubin excretion in Man and the Rat. Birth Defects: Original Article Series 12:81 (1976)

19. A. F. McDonagh and L. Ramonas. Jaundice Phototherapy: Micro Flo-Cell Photometry Reveals Rapid Biliary Response of Gunn Rats to Light. Science 201:829 (1978)

20. T. P. Vogl, H. Cheskin, T. A. Blumenfeld, W. T. Speck, and R. Koenigsberger. Effect of Intermittent Phototherapy on Bilirubin Dynamics in the Gunn Rat. Ped. Res. 11:1021 (1977)

21. T. Vogl. Bilirubin Redistribution in the Skin of a Crigler-Najjar Child Under Phototherapy. Paper P218 Presented at the VII International Congress on Photobiology, Rome, Italy, September 2, 1976

22. T. P. Vogl. Phototherapy of Neonatal Hyperbilirubinemia: Bilirubin in Unexposed Areas of the Skin. J. Pediatr. 85:707(1974)

23. A. F. McDonagh, D. A. Lightner, and A. Wooldridge. Geometric Isomerization of Bilirubin and its Dimethyl Ester. J.C.S. Chem. Comm. p.110 (1979)

24. R. Schmid. Bilirubin Metabolism: State of the Art. Gastroent. 74:1307 (1978)

25. R. Brodersen. Bilirubin: Solubility and Interaction with Albumin and Phospholipid. J. Biol. Chem. 254:2364 (1979)

26. R. E. Behrman, A. K. Brown, M. R. Currie, L. C. Harber, J. W. Hastings G. B. Odell, R. Schaffer, R. B. Setlow, T. P. Vogl, and J. Wurtman. Final Report of the Committee on Phototherapy in the Newborn. National Academy of Sciences, Washington, D.C. 1974

27. R. Schmid and L. Hammaker. Metabolism and Disposition of C^{14}-Bilirubin in Congenital Nonhemolytic Jaundice. J. Clin. Invest. 42:1720 (1963)

28. J. D. Ostrow. Photocatabolism of Labeled Bilirubin in the Congenitally Jaundiced (Gunn) Rat. J. Clin. Invest. 50:707 (1971)

29. T. P. Vogl, T. Hegyi, I. M. Hiatt, R. A. Polin, and L. Indyk. Intermittent Phototherapy in the Treatment of Jaundice in the Premature Infant. J. Pediatr. 92:624(1978)

30. T. P. Vogl, H. Cheskin, T. A. Blumenfeld, and K. Shugai. Bilirubin Dynamics in the Gunn Rat: Dose Response of Continuous and Intermittent Phototherapy. Submitted for publication

31. R.F. Brown and K.R. Godfrey. Problems of Determinacy in Compartmental Modeling with Application to Bilirubin Kinetics. Math. Bioscien. 40:205 (1978)

32. E. R. Carson and E.A. Jones. Use of Kinetic Analysis and Mathematical Modeling in the Study of Metabolic Pathways In Vivo. NEJM 300:1016 & 1078 (1978)

33. R. F. Brown, E.R. Carson, L. Finkelstein, K.R. Godfrey, and P.R. Jones. Adequacy of Measurements in Compartmental Modeling of Metabolic Systems. Med. & Biol. Eng. & Comp. 17: 216 (1979)

34. L. Ballowitz, G. Gentler, J. Krochman, R. Pannitschka, G. Roemer, and I. Roemer. Phototherapy in Gunn Rats. Biol. Neonate 31:229 (1977)

35. J. F. Lucy. Neonatal Jaundice and Phototherapy. Ped. Clin. N. A. 19:827 (1972)

36. A.E. Kopelman, R.S. Brown, G.B. Odell. The "Bronze" baby syndrome: A Complication of Phototherapy. J. Pediatr. 81:466 (1972)

37. H. M. Maurer, M. Fratkin, N.B. McWilliams, B. Kirkpatrick, D. Draper, J.C. Higgins, C.R. Hunter. Effects of Phototherapy on Platelet Counts in Low Birth Weight Infants and on Platelet Production and Life Span in Rabbits. Pediatr. 57:506 (1976)

38. H. M. Maurer, J.C. Higgins, W.J. Still. Platelet Injury During Phototherapy Amer. J. Hemat. 1:89 (1976)

39. D. Bratlid. The Effects of Antimicrobial Agents on Bilirubin Binding by Human Erythrocytes. Scan. J. Clin. Lab. Invest. 30:331(1972)

40. W.T. Speck, H.S. Rosenkranz. Intracellular DNA-modifying Activity of Phototherapy Lights. Ped. Res. 10:553 (1976)

41. R. M. Santella, H. S. Rosenkranz, W.T. Speck. Intracellular DNA-modifying Activity of Intermittent Phototherapy. J. Ped. 93:106 (1978)

42. C.E. Aplin, B.H. Brouhard, R.J. Cunningham, J. Richardson. Phototherapy and Plasma Immunoreactive Prostaglandin A values. AM. J. Dis. Child. 133:625 (1979)

43. P. Y. K. Wu and J. E. Hodgeman. Insensible Water Loss in Preterm Infants: Changes with Postnatal Development and Non-ionizing Radiation. Pediatrics, 54:704 (1974)

44. E.F. Bell, G.A. Neidich, W.J. Cashore, W. Oh. Combined Effect of Radiant Warmer and Phototherapy on Insensible Water Loss in Low-Birth-Weight Infants. J. Ped. 94: 810 (1979)

45. R. M. Klein. Shedding Light on the use of Light, Pediatr. 50:118 (1972)

46. L. C. Mims, M. Estrada, D. S. Gooden, R. R. Caldwell, and R. V. Kotas. Phototherapy for Neonatal Hyperbilirubinemia - A Dose : Response Relationship. J. Pediatr. 83:658 (1973)

47. K. L. Tan. The Nature of the Dose-Response Relationship of Phototherapy for Neonatal Hyperbilirubinemia. J. Ped. 90: 448 (1978)

*   Note added in proof: Contributing to this unravelling of the configurational changes at the C5 and C15 bridges of bilirubin was X-ray crystallographic analysis of bilirubin performed by Bonnet, Davis, and Hursthouse [Nature 262:326 (1976)]. This group of investigators independently also proposed photoisomerization as the mechanism of phototherapy [Proc. Roy. Soc., London, B. 202:249 (1978)].

Photoinduced Modification of Human Serum Albumin During Phototherapy of Jaundiced Newborns

G. Jori, F. Rubaltelli[+], and E. Rossi

Istituto di Biologia Animale, Centro C.N.R. Emocianine, Università di Padova
and [+]Clinica Pediatrica, Università di Padova
Padova, Italy

1. Introduction

Phototherapy with visible light is a world-wide accepted method for the treatment of neonatal unconjugated hyperbilirubinemia. Even if the phototreatment may be accompanied by the onset of side effects (see Table 1), the inherent advantages overcome the possible risks. The mortality rate, associated with exchange transfucion, i.e. with the major alternative treatment for hyperbilirubinemia, is about 1 percent in the leading neonatal units.

Table 1 Possible side effects of phototherapy

Observed effect	Reference
Decreased gut transit time, probably related to lactase deficiency	1, 2
Abnormal urinary excretion of tryptophan metabolites	3
Reduced erythrocyte level of glutathione	4
Reduced blood level of riboflavin	5
Increased peripheral blood flow and insensible water loss	6
Reduced platelet counts	7
Bronze baby syndrome	8
Hypocalcemia	9

Several lines of research are presently active to clarify the molecular mechanisms of bilirubin photodegradation. A major goal of such investigations is the quantitative definition of dose – effect relationships to avoid overexposure of the infants to the incident light. A possible way to minimize the side effects might be a strict control of the total amount of light energy supplied and/or of the range of light wavelengths used for irradiation.

2. Bilirubin-sensitized photochemical modification of biomolecules "in vitro"

Several biological systems are photomodified by blue light-irradiation in the presence of bilirubin (Table 2). Nearly all classes of cell constituents are affected, including proteins, nucleic acids, lipids and membranes. On the basis of the photodamaged sites, the photosensitizing action of bilirubin probably involves the intermediacy of radical species, generated (at least in part) through electron ejection from the photoexcited tetrapyrrole (19). Potentially competing photoreaction mechanisms, involving attack on the various substrates by activated oxygen species, originated by energy transfer from triplet bilirubin, should play a minor role: photoexcited bilirubin is a poor generator of singlet oxygen and displays a high reactivity toward the latter species (20, 21).

This conclusion is further supported by the finding that some bilirubin-sensitized photoreactions (18, 22) proceed smoothly in degassed solutions.

Table 2   Bilirubin-sensitized photomodification of biological systems

Substrate	Observed changes	Ref.
N-acetyl-L-cysteine Reduced glutathione	Formation of bilirubin-substrate photoadducts	10, 11
DNA	Cleavage of phosphodiester bonds	12
Human erythrocyte membranes	Cross-linking of proteins in the inner face of membrane	13
Neonatal erythrocytes	Hemolysis due to K^+ leakage and oxidation of glutathione and/or lecithin	14-16
Human plasma	Destruction of some amino acids (e.g. tryptophan and cysteine)	17
Human and bovine serum albumin	Oxidation of histidyl and tryptophyl side chains close to the bilirubin binding site	18

3. Irradiation of the human serum albumin-bilirubin complex

Albumin is a likely target for bilirubin-sensitized photoeffects "in vivo", since this protein is the natural carrier of bilirubin in the serum. We undertook a detailed study on the irradiation of the 1:1 complex between bilirubin and human serum albumin.

A 0.1 mM solution of the complex in 0.5 M phosphate buffer, pH 7.4, was exposed to the light of a 1250 W halogen lamp, both in air-equilibrated and in deaerated solutions. A chemical filter was placed in the front of the cuvette containing the irradiated solution to intercept wavelengths below 330 nm. The irradiance on the bilirubin-albumin system in the 440-470 nm range was 300 $\mu W/cm^2/nm$.

The protein-bound bilirubin was photodegraded with a slower kinetics than free bilirubin, suggesting some interaction between albumin and the photoexcited dye. One explanation might involve the competition between bilirubin and some amino acid residues of albumin for the photosensitizing agents. As shown in Table 3, amino acid analyses of albumin complexes, after different irradiation times, pointed out that some histidyl residues were preferentially modified. The affected histidines are probably present at the albumin binding site for bilirubin (18).

Table 3 Alteration of chemical and functional properties of albumin upon irradiation of its 1:1 complex with bilirubin

---

1. Preferential photomodification of two histidyl and of the single tryptophyl residue. Slower photomodification of tyrosyl residues.

2. Decrease of the bilirubin binding constant from 2.07 to 0.54x 10^8 M^{-1}.

3. Formation of a covalent photoadduct between bilirubin and the protein. The photoadduct specifically occurs in the section of albumin molecule between residues 187 and 397; its yield is higher upon irradiation in deaerated solutions.

4. Decreased resistance to denaturing agents, such as temperature and guanidinium chloride.

The chemical alteration of selected amino acid side chains exerted a remarkable influence on several functional and conformational properties of serum albumin (Table 3). On the whole, the observed changes indicated that the albumin modified at two histidines had a reduced affinity for bilirubin; after the additional photomodification of tyrosines, albumin displayed also a three-dimensional organization less compact than that typical of the native protein.

Interestingly, after irradiation, bilirubin or some photoproduct could not be removed completely from albumin, even after
unfolding of the protein by precipitation with acetone and/or incubation in the presence of the denaturant 7 M guanidine. The
non-removable fraction thus appeared to be covalently bound with
albumin. Cleavage of irradiated bovine serum albumin with cyano-
gen bromide and subsequent column chromatographic separation of
the fragments yielded a single bilirubin-containing peptide, which
was identified as peptide 187-397 of native bovine serum albumin
(23). Therefore, the photoaddition process appears to be characterized by a high spatial selectivity.

4. "In vivo" formation of a bilirubin-albumin photoadduct

To evaluate the relevance of the described findings for the phototherapy of jaundiced newborns, we investigated whether the
photoaddition between bilirubin and albumin is occurring also
"in vivo". The sera of 29 full-term hyperbilirubinemic newborns
were analyzed as shown in Fig. 1. The procedure, which allows the
isolation of the serum albumin fraction, was applied at different
times before, during and after the phototreatment. The irradiance
at the body level was about 22 $\mu W/cm^2/nm$.

PROCEDURE FOR ANALYSIS OF SERA OF JAUNDICED
INFANTS.

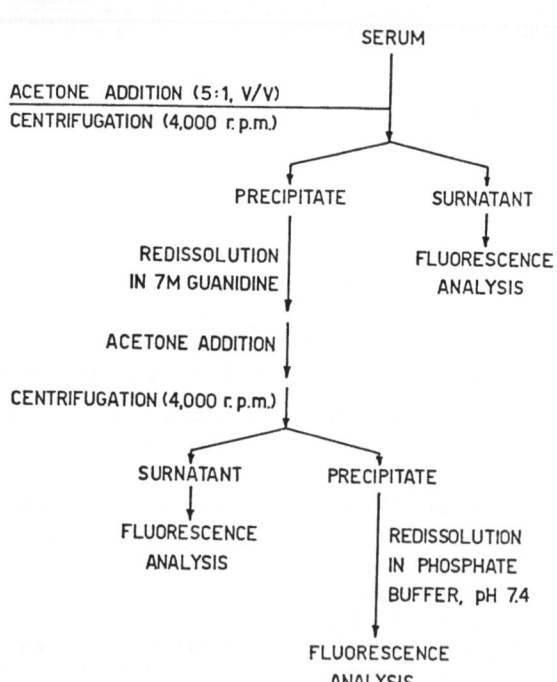

Fig. 1 Analytical
procedure adopted
for the isolation
of the albumin
fraction from the
sera of jaundiced
newborn infants

148

The photoadduct formation was followed by spectrophotofluorimetric analysis, taking advantage of the characteristic fluorescence excitation (Fig. 2) and emission (Fig. 3) spectra of albumin-bound bilirubin; as shown in the Figures, under our experimental conditions, no appreciable amount of albumin-bound fluorescent material was found in unirradiated samples. On the other
hand, both excitation and emission spectra typical of bilirubin
were obtained after bilirubin-albumin isolation from the sera of
phototreated patients.

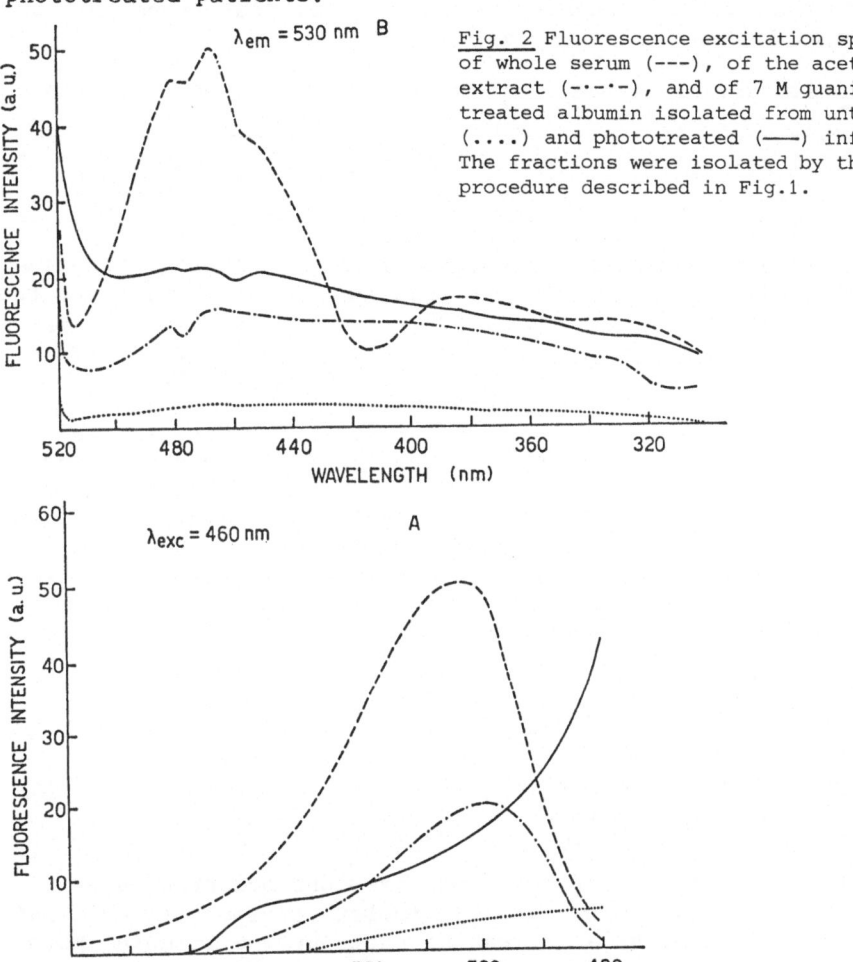

Fig. 2 Fluorescence excitation spectra
of whole serum (---), of the acetone
extract (-·-·-), and of 7 M guanidine-
treated albumin isolated from untreated
(....) and phototreated (——) infants.
The fractions were isolated by the
procedure described in Fig.1.

Fig. 3 Fluorescence emission spectra of whole serum (---), of the
acetone extract (-.-.-), and of 7 M guanidine-treated albumin i-
solated from untreated (.....) and phototreated (——) infants.
The fractions were isolated by the procedure described in Fig. 1.

149

The set of data, obtained after consideration of the clinical
history of each patient, is summarized in Table 4. In all cases,
the rate of formation of the photoadduct increased upon increasing
the intensity of the incident light. The photoadduct disappeared
about 15-20 days after the interruption of the phototreatment:
this period of time represents the natural turn-over period of
serum albumin. Therefore, the photoaddition process must be of
general occurrence, although its apparently low yield and transient presence in the organism would indicate that there are only minor biological consequences.

Table 4 Summary of clinical studies with phototreated infants

Number of investigated subjects: 29	Schedule of the phototreatment	Observed phenomena
Serum bilirubin levels at the beginning of the phototreatment: 10.8 - 15 mg/dl	Light source: four F20T12/BB Westing-house lamps	Average irradia-tion time for the photoadduct ap-pearance: 7-9 hours.
	Irradiance in the 440-470 nm range: 22 µW/sq. cm/nm	Average time for photoadduct disappearance:
	Type of treatment: continuous irradiation	15-20 days

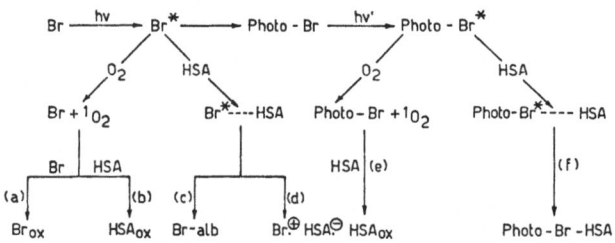

Fig. 4 Proposed scheme for the photoreactions occurring upon visible light-irradiation of the 1:1 complex between bilirubin and
human serum albumin. Br = bilirubin; Photo-Br = bilirubin photoisomer; HSA = human serum albumin.

As yet, the photochemical mechanism(s) underlying the photoadduct formation have not been completely elucidated. On the basis of our data and of literature reports on bilirubin photochemistry (24, 25), the tentative scheme outlined in Fig. 4 can be
proposed.

The scheme depicted in Fig. 4 emphasizes the possibility of energy transfer from photoexcited bilirubin or its photoexcited photoisomers to molecular ground state oxygen with generation of singlet oxygen. As a consequence, bilirubin photooxidation – process (a) – or albumin photooxidation – processes (b) and (e) – may take place: no photooxidation of the photoisomers has been considered, since these compounds are eliminated as such (26). However, a more important photoreaction pathway should involve the direct interaction between the protein and photoexcited Br or Photo-Br. The photoadduct formation may proceed by sequences (c) or (f). The yield of these steps is limited by the possible intervention of electron transfer phenomena – process (d) – e.g. leading to a bilirubin radical cation. The radical species thus formed may undergo further reactions (e.g., chain processes), causing the irreversible alteration of other substrates.

5. Conclusions

Our findings demonstrate that the lack of complete bilirubin removal from albumin after "in vitro" visible light-irradiation is due to the formation of a covalent protein-pigment photoadduct. The same phenomenon occurs "in vivo" pointing out that phototherapy does induce some kind of photodamage. The "in vitro" photoaddition lowers the bilirubin-albumin affinity constant. No similar observation was reported after "in vivo" irradiation; probably, the amount of photoaltered albumin is small, since the photoadduct was barely detectable by absorption spectroscopy, a less sensitive technique than spectrophotofluorimetry. Also, the methods clinically available for estimating albumin binding capacity may lack sufficient sensitivity to detect small changes of the parameter. The demonstration of the actual occurrence of "in vivo" photodamage stresses the importance of studies aimed at avoiding overexposure of patients to the light source.

6. References

1. Rubaltelli F.F. and Largajolli G., 1973, Acta Paediat. Scand. 62, 146.
2. Bakken A.F., 1976, Pediat. Res. 10, 186.
3. Rubaltelli F.F., Allegri G., Costa G. and De Antoni A., 1974, J. Pediat. 85, 865.
4. Blackburn M.G., Orzalesi M. and Pigram P., 1972, J. Pediat. 80, 640.
5. Sisson T.C.R., Slaven B. and Hamilton P.B., 1976, Natl. Found. Birth Defcts: Orig. Art. Ser. 13, 122.
6. Oh W., Williams P.R., Tao A.C. and Lind J., 1976, Natl. Found. Birth Defects: Orig. Art. Ser. 13, 114.

7. Maurer H.M., McWilliams N.B. and Hunter C., 1975, Pediat. Res. 9, 368.
8. Kopelman A.E., Brown R.S. and Odell G.B., 1971, Pediat. Res. 5, 642.
9. Romagnoli C., Polidori G., Catldi L., Tortorolo G. and Segni G., 1979, J. Pediat. 94, 815.
10. Manitto P., Monti D. and Garbagnati E., 1972, Il Farmaco, Ed. Sci., 27, 999.
11. Manitto P. and Monti D., 1976, J. Chem. Soc. Chem. Commun. 122-123.
12. Speck T.W., 1974, Pediat. Res. 8, 451.
13. Girotti A.W., 1975, Biochemistry 14, 3377.
14. Blackburn M.G., Orzalesi M. and Pigram P., 1972, Biol. Neo-nate 21, 35.
15. Odell G.B., Brown R.S. and Kopelman A.L., 1972, J. Pediat. 81, 473.
16. Castro M., Tambucci S., Panero A., Giardini O. and Orzalesi, M., 1976, Min. Pediat. 28, 391.
17. Engelhardt D.L., Santella R.M., Rosenkranz H.S. and Speck W.T., 1977, Photochem. Photobiol. 26, 53.
18. Rubaltelli F.F. and Jori G., 1979, Photochem. Photobiol. 29, 991.
19. Lightner D.A., Woolridge T.A. and McDonagh A.F., 1979, Bio-chem. Biophys. Res. Commun. 86, 235.
20. McDonagh A.F., 1974, Phototherapy of the Newborn: an Overview, Natl. Acad. of Sciences, Washington, pp. 56-73.
21. Foote C.S. and Ching T.Y., 1975, J. Am. Chem. Soc. 97, 6209.
22. Lightner D.A., 1977, Photochem. Photobiol. 26, 427.
23. King T.P. and Spencer M., 1970, J. Biol. Chem. 245, 6134.
24. Lightner D.A., Landen G.L. and Rodgers S.L., 1979, Abstarcts 7th ASP meeting, Asilomar, California, p. 59.
25. Rubaltelli F.F., Jori G. and Rossi E., 1979, The Lancet, sub-mitted for publication.
26. Lightner D.A., Woolridge T.A. and McDonagh A.F., 1979, Proc. Natl. Acad. Sci. U.S. 76, 29.

A Laser Flash Photolysis Study of Bilirubin

R.S. Sinclair, R.W. Sloper, and T.G. Truscott

Department of Chemistry, Paisley College of Technology
Paisley, Renfrewshire PA1 2BE, Scotland

1. Introduction

Bilirubin (BR) which is known to bind strongly to human serum albumin (HSA) [1,2,3] and to lipids, lipoproteins and cell membranes [4,5] can be considered, in the body to be distributed according to the equilibrium:

$$BR - HSA \; \rightleftharpoons \; BR \; (free) \; \rightleftharpoons \; BR - tissue$$

The high BR level in infants suffering from neonatal jaundice exposes them to the risk of kernicterus in which irreversible tissue damage results from the uptake of BR by brain cells [6]. Although phototherapy, in recent years, has proved extremely effective in the reduction of BR levels in jaundiced infants, the precise mechanism for this process is still not fully understood and has provoked much photochemical study [7]. One interpretation postulates self sensitised photooxidation [8] and another proposes photoisomerisation [9,10] of BR.

An understanding of the photophysics of BR is important for the elucidation of the mechanism of phototherapy. Self sensitised photooxidation requires that the excited singlet state, formed by optical excitation crosses over to the triplet state which undergoes energy transfer with molecular oxygen resulting in the formation of singlet oxygen. Although direct evidence for the participation of triplet BR is somewhat scant an upper limit of 0.1 [11] has been set for its quantum yield of formation (Φ_T) which is high enough to account for a photooxidation mechanism.

The shortness of duration of Q-switched laser pulses renders laser techniques particularly suitable for the study of fast processes involving species with lifetimes $\leqslant 10$ μs which arise following optical excitation of molecules. We report here results from the 347 nm frequency doubled ruby laser flash photolysis of BR in a variety of environments including its complexes with HSA.

2. Experimental

BR IX α was supplied by Sigma and HSA, Cohn fraction V by Fluka. Each was used without further purification although all solvents were extensively purified. The laser flash photolysis apparatus has previously been described [12]. Quantum yields were measured by the comparative method [13,14] with anthracene in cyclohexane as actinometer, taking for anthracene $\Phi_T = 0.71$ and $\varepsilon_T^{422} = 6.47 \times 10^4 dm^3 mol^{-1} cm^{-1}$ [15,16] and for BR, $\varepsilon_T^{500} = 8.8 \times 10^3 dm^3 mol^{-1} cm^{-1}$ [11].

3. BR in Various Solvent Systems

Samples of BR in polar and non polar, basic and neutral solvents were subjected to 347 nm laser flash photolysis using pulses of sufficiently low intensity to avoid anomalous transients from the solvents. Under these conditions no transients were observable in the 400-600 nm range with water (pH 9-11), 10% aqueous mulgofen, 2.2 g dm^{-3} aqueous cetyl trimethylammonium bromide or with chloroform containing 0-20% triethylamine. Comparison with anthracene enabled us to set ϕ_T < 0.005 in these cases.

In contrast to other work [17] we have not been able to detect triplet BR in 1% methanolic ammonia following 347 nm excitation. A permanent product with a spectrum resembling, in the longer wavelength region, those attributed by McDONAGH [10] to photoisomers was produced within 200 ns of the laser pulse. Due to the very high optical density of the ground state we were unable to measure the shorter wavelength part of the spectrum by laser flash photolysis. This species was unaffected by oxygen but was slowly decomposed by the monitoring lamp (λ > 490 nm) consistent with McDONAGH's observations.

In benzene (OD_{347} = 0.23) under the same conditions two weakly absorbing species ($\Delta OD \sim 2 \times 10^{-3}$) were observed: a product with maximum absorption and depletion at 550 and 500 nm respectively together with an oxygen quenchable transient of lifetime about 10 µs peaking at 520 nm. The product spectrum, which may be due to a photoisomer is qualitatively similar to differential absorption spectra taken after continuous irradiation with blue light (410 - 470 nm) but differs in the positions of the absorption and depletion maxima (495 and 465 nm respectively). The transient, which can be identified as triplet BR, was estimated by anthracene actinometry to have $\phi_T \simeq 0.01$.

4. BR Complexed with HSA

347 nm laser flash photolysis of 8.3×10^{-5}M BR and 1.2×10^{-4}M HSA in phosphate buffered water at pH 7.40 yielded two species: a short lived transient which decayed by first order kinetics with k = $3.5 \times 10^5 s^{-1}$ ($\tau \simeq 3$ µs) and a permanent product which was slowly destroyed by the monitoring lamp (λ > 490 nm). Neither species was affected by O_2, N_2O or I^- and variation of laser intensity showed both species to be formed by monophotonic processes. Spectra of both species are presented in Fig. 1 together with a differential absorption spectrum measured after irradiation of a sample for 30 sec. with blue light (410 - 470 nm). The spectra are again all similar in the longer wavelength region (the loss peak was not observable by laser flash photolysis) and also similar to spectra attributed to photoisomers [10].

In an attempt to determine which of the HSA binding sites was involved in the production of these species, transient and product optical densities were plotted as a function of concentration in Fig. 2. The shape of the curves strongly suggests that most if not all of the transient and product originates from BR molecules occupying the first, most strongly binding site in the protein. That a single site is involved is further supported by the constancy of the transient decay constants over the entire concentration range. The non-linearity of Fig. 2 and our failure to produce these species by exciting the protein absorption band with the 265 nm laser shows that the transient and product are not the result of energy transfer from the protein.

The yields of both species and the decay of the transient were found to be strongly temperature dependent, an increased temperature resulting in lower

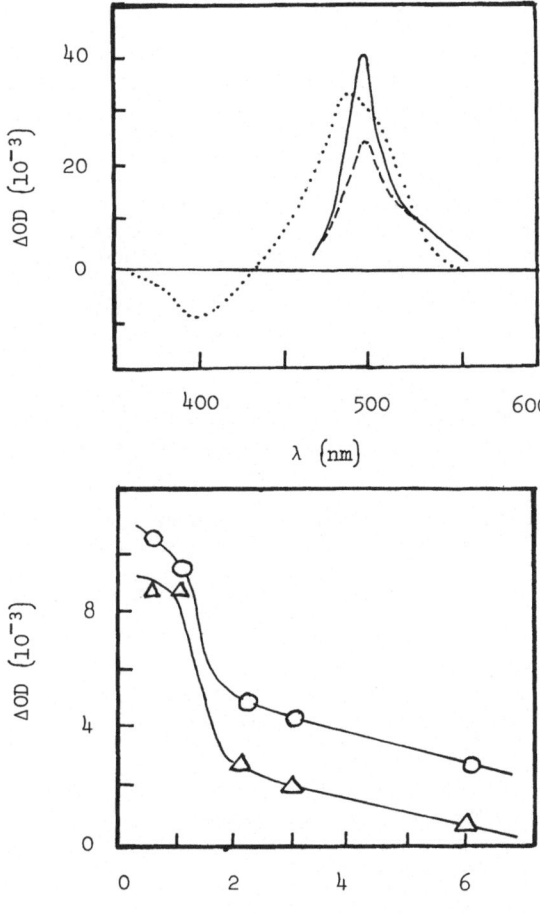

Fig.1 Differential spectrum (······) and laser flash photolysis spectra of transient (- - -) and product (———) following 347 nm laser flash photolysis of BR in HSA

Fig.2 Variation of transient (0) and product (Δ) yields with composition

yields and faster decays corresponding to an activation energy of about 50–60 kJ mol^{-1}. Although the product can be assigned to a mixture of ZE and EZ photoisomers (10), identification of the transient is a little more difficult. A possible explanation is that the transient is an excited state of BR which twists to form a photoisomer. If BR, which is known (18,19) to occupy a hydrophobic cleft in the protein, is internally hydrogen bonded this explanation is energetically consistent with our results. The observed activation energy would be made up from contributions of 38 kJ mol^{-1} for the rupture of three hydrogen bonds (20) and about 10–20 kJ mol^{-1} for an excited state rotation (measured from diagrams published by FALK (21)). The excited state involved is presumably a triplet in view of its 3 μs lifetime but it is rather strange that it is so insensitive to the presence of oxygen. It could perhaps be that the protein very effectively protects the BR from exposure to oxygen.

5. Conclusions

Although the yields of triplet following optical excitation of BR in various solvents have been shown to be extremely small the photoisomerisation process has in certain cases been shown to occur very rapidly (within 200 ns of the laser pulse in methanolic ammonia). In HSA the photoisomerisation process is also fast. If the transient is a precursor to photoisomers then these are formed with a time for half reaction of about 3 μs. If however the transient is not a precursor their formation may be even more rapid. Since light of wavelengths used in phototherapy is known to penetrate the skin, photoisomerisation of BR in the plasma may make an important contribution to the phototherapeutic process.

References

1. Ostrow, J.D. and R. Schmid, J. Clin. Invest. (1963), 42, 1286.
2. Beaven, G.H., A. d'Albis and W. Gratzer, Eur. J. Biochem. (1973), 33, 500.
3. Chen, R.F., in "Biological Fluorescence Concepts", ed. A.A. Thaer and M. Sernetz, Springer Verlag (1973), 273.
4. Mustafa, M.G. and T.E. King, J. Biol. Chem. (1970), 245, 1084.
5. Cowger, Marilyn L., in "Phototherapy of the Newborn – an overview", ed. G.B. Odell, R. Schaffer and A.P. Simopoulos, Nat. Acad. Sci. U.S.A. (1974), 93.
6. Diamond, I., in Adv. Pediatr., ed. I. Schulman, Chicago Year Book Medical Publishers (1969), 16, 99.
7. Lightner, D.A., Photochem. Photobiol. (1977), 26, 427.
8. McDonagh, A.F., in "Bilirubin Metabolism in the Newborn", ed. D. Bergsma and S.H. Blondheim, National Foundation March of Dimes (1976), 40.
9. Lightner, D.A. and Y.T. Park, Tetrahedron Lett. (1976) 2209.
10. McDonagh, A.F., D.A. Lightner and T.A. Wooldridge, J.C.S., Chem. Commun. (1979) 110.
11. Land, E.J., Photochem. Photobiol. (1976), 24, 475.
12. McVie, J., R.S. Sinclair and T.G. Truscott, J.C.S., Faraday II (1978), 74, 1870.
13. Richards, J.T. and J.K. Thomas, Trans. Faraday Soc. (1970), 66, 621.
14. Bensasson, R.V., C.R. Goldschmidt, E.J. Land and T.G. Truscott, Photochem. Photobiol. (1978), 28, 277.
15. Birks, J.B., ed. "Organic Molecular Photophysics" Vol.I, John Wiley, 1973.
16. Birks, J.B., ed. "Organic Molecular Photophysics" Vol.II, John Wiley, 1974.
17. Barber, D.J.W. and J.T. Richards, Chem. Phys. Lett. (1977), 46, 130.
18. Krasner, J. and S.J. Yaffe, in "Bilirubin Metabolism in the Newborn II", ed. D. Bergsma and S.H. Blondheim, National Foundation March of Dimes (1976), 168.
19. Kuenzle, C.C., N. Cumarasamy and K. Wilson, in "Metabolism and Chemistry of Bilirubin and Related Tetrapyrolles", ed. A.F. Bakken and J. Fog, Pediatr. Res. Inst., Rikshospitalet, Oslow (1975), 39.
20. Manitto, P. and D. Monti, J.C.S., Chem. Commun. (1976), 122.
21. Falk, H., K. Grubmayr and T. Schlederer, Monatshefte für Chemie (1978), 109, 1191.

Part V

Absorption and Fluorescence Spectroscopy

Laser Fluorescence Spectroscopy of Biomolecules

A. Anders

Institut für Biophysik, Universität Hannover
D-3000 Hannover 1, Fed. Rep. of Germany

In the last years lasers, especially tunable dye lasers [1] became indispensable tools in all fields of spectroscopy because of their great advantage over conventional light sources [2, 3] . They gained growing importance not only in photophysics and photochemistry [3, 4] but recently in photobiology and photomedicine, too [4, 5, 6] . Laser investigations will increase our knowledge about structures and spectroscopic properties of biomolecules as well as about kinetics of various biochemical processes.

Spectroscopy of biomolecules covers a wide field of investigations. This review concentrates on topics in fluorescence spectroscopy. Other subjects like Raman, absorption, picosecond and excited-state spectroscopy are discussed in separate papers of this book. Further spectroscopic applications are described in the literature, e.g. Doppler spectroscopy [7] , and photoreactivation processes [8] .

This paper is mainly concerned with nucleic acids. Other biomolecules like chlorophyll, rhodopsin, myo- and hemoglobin are reviewed in the literature [9, 1o, 11] .

In the beginning, those features of lasers are considered which are useful to investigate biomolecules, followed by some typical applications:

- high intensity + mono-
  chromaticity (i.e. high
  spectral intensity)

  fluorescence; multi-photon processes;
  excited states; absorption of biological
  material

- tunability

  selective excitation

- short pulses

  lifetimes; fast reactions

- small focus

  irradiation of selected parts,
  e.g., in cells or tissue

High intensity and monochromaticity - resulting in a high spectral intensity - allow fluorescence investigations with low quantum yields and the study of multi-photon processes and excited states (reaching a high population), for example. Another application is the measurement of absorption and transmission spectra of thick specimen like human skin. One can perform selective excitations of single molecular states or of distinct molecules in a mixture using the additional feature of a dye laser, the tunability, together with the high spectral intensity. The generation of tunable UV radiation is of special importance, because many electronic resonances of biologically interesting molecules (like nucleic acids and proteins) are situated in the UV region. Short pulses give the possibility of observing lifetimes or fast reactions of molecules in the picosecond time scale. Furthermore, the small focus (in the order of 1 μ) enables irradiation of selected parts of cells or tissue.

1. Experimental Techniques

Laser investigations of biomolecules depend on the development of laser technology and on properties like intensities, bandwidth, pulse and continuous wave operation in the different wavelength regions from UV to IR [1, 10, 12, 13].

Molecules like chlorophyll and rhodopsin can be investigated with visible excitation because of their visible absorption bands. The lasers mainly used are tunable dye lasers (about 360 - 900 nm) and the fixed frequency lasers: HeNe (628 nm), Ar⁺ (different blue and green lines), Ruby (694 nm), Nd:Yag (532 nm, second harmonic wavelength). But other biomolecules, like nucleic acids and proteins, absorb only in the UV region. In order to investigate them frequency doubled dye lasers (down to 220 nm), N₂-lasers (337 nm) and Nd:Yag-lasers (266 nm, fourth harmonic wavelength) were applied. Furthermore, the stimulated Raman scattering in hydrogen gas excited by a ND:Yag pumped dye laser generates tunable radiation from 185 to 880 nm [12]. Another method of studying biomolecules absorbing in the UV is the use of special dyes attached to the biomolecules and investigating the visible dye fluorescence (secondary fluorescence). This method is also suitable to obtain information about molecules having a very low quantum yield of fluorescence.

A typical arrangement for the observation of fluorescence in solution after excitation with a pulsed dye laser (in the visible and UV-region) is given in Fig. 1. The conditions should closely resemble the situation in living cells (pH 7 etc.).

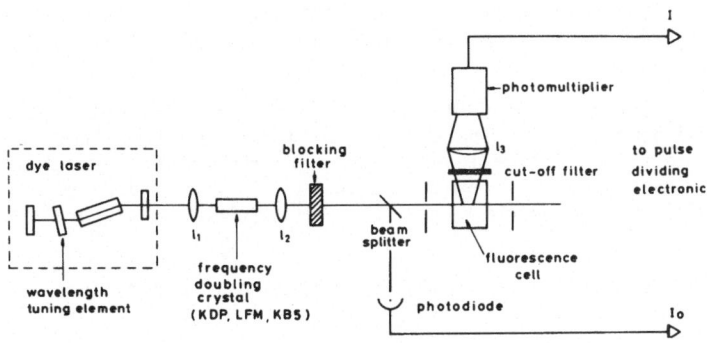

Fig. 1 Experimental arrangement to measure the fluorescence of biomolecules in solution after UV excitation with a pulsed dye laser [21]

The excitation of special parts in cells is performed with microbeam systems, where the laser beam is coupled into a microscope [14,15]. A typical set-up is shown in Fig. 2. In this case a nitrogen laser pumped dye laser is used.

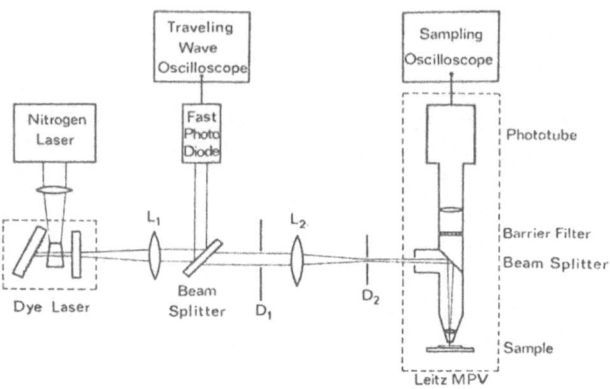

Fig. 2   Microbeam system for fluorescence observations (see text) [16]

2. Nucleic Acids

Nucleic acids store and transmit genetic information, and regulate metabolism in living cells. The genetic code is represented in the sequence of bases in the DNA.

The measurement of fluorescence of nucleic acids at room temperature and neutral pH belongs to those problems which involve extremely low fluorescence quantum yields in the order of 10^{-4} to 10^{-5}. Processes like energy migration along the DNA chain and the deactivation of the electronic states are expected to be on a picosecond time scale. Picosecond applications in DNA are described by CAMPILLO et al.[11].

Because of the low quantum yield of fluorescence nearly all investigations were done at low temperatures (77K). Therefore, the fluorescence spectrum of DNA at room temperature is not yet well understood. The fluorescence spectrum of calf thymus DNA was measured after excitation with a frequency doubled pulsed dye laser at 266 nm, the quantum yield at room temperature was evaluated to about $2 \cdot 10^{-5}$ [17] .

Energy migration processes along the DNA may be studied in DNA-dye complexes (see chapter 3).

Information about conformation and motion of nucleic acids was obtained by EHRENBERG et al. [18] using pulsed fluorescence and fluorescence correlation spectroscopy. They investigated transfer-RNA which was labelled with the dye ethidium bromid. Transfer-RNA molecules are participated in protein synthesis at the ribosomes. The results indicate that conformational changes of the transfer-RNA molecules are important for the interaction between codon and anticodon at the ribosomes, for example.

3. Nucleic Acid-Dye Complexes

The interaction of biologically active molecules like dyes with nucleic acids is of great interest, because such molecules are used as drugs in chemotherapy or can induce mutations and tumors, for example. They form intermolecular complexes with nucleic acids. A special class is considered those dyes which interact with DNA by intercalation (Fig. 3) [19,20].

A base-sequence dependent effect manifests in a different fluorescence behaviour or energy transfer if dyes are intercalated between different base pairs. For the investigation of such effects results of DNA of different base compositions are compared [20,21].

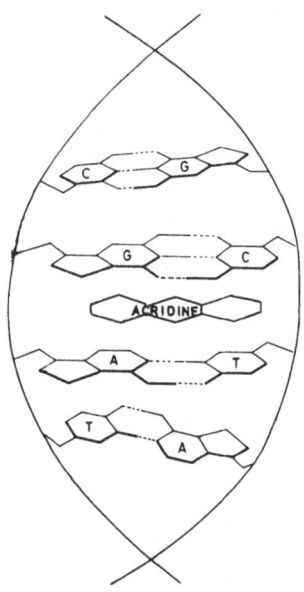

Fig. 3 Intercalation principle: Insertion of a planar dye molecule between adjacent base pairs (called base-pair unit) in the DNA double helix. Three combinations are possible: insertion between two adenine-thymine (AT) or guanine-cytosine (GC) pairs or an AT and a GC base pair sequence

First, one example is described, where the fluorescence lifetime is base-sequence-dependent. As the lifetimes of the excited states of nucleic acids are in the ps range, short pulse techniques are needed for time resolved observations. The double base pair content (e.g. AT-AT) in chromosomes can be determined comparing the fluorescence lifetimes of chromosomes stained with dyes and those of DNA-dye solutions. After staining some chromosome parts, called bands, fluoresce more intensively than others (Fig. 4 b). This phenomenon is used for chromosome characterization. Fig. 4 a shows the "in band" and "out of band" fluorescence of a Vicia Faba chromosome stained with quinacrine mustard. The results seem to indicate that the "in band" parts contain more AT-AT sequences[19, 22].

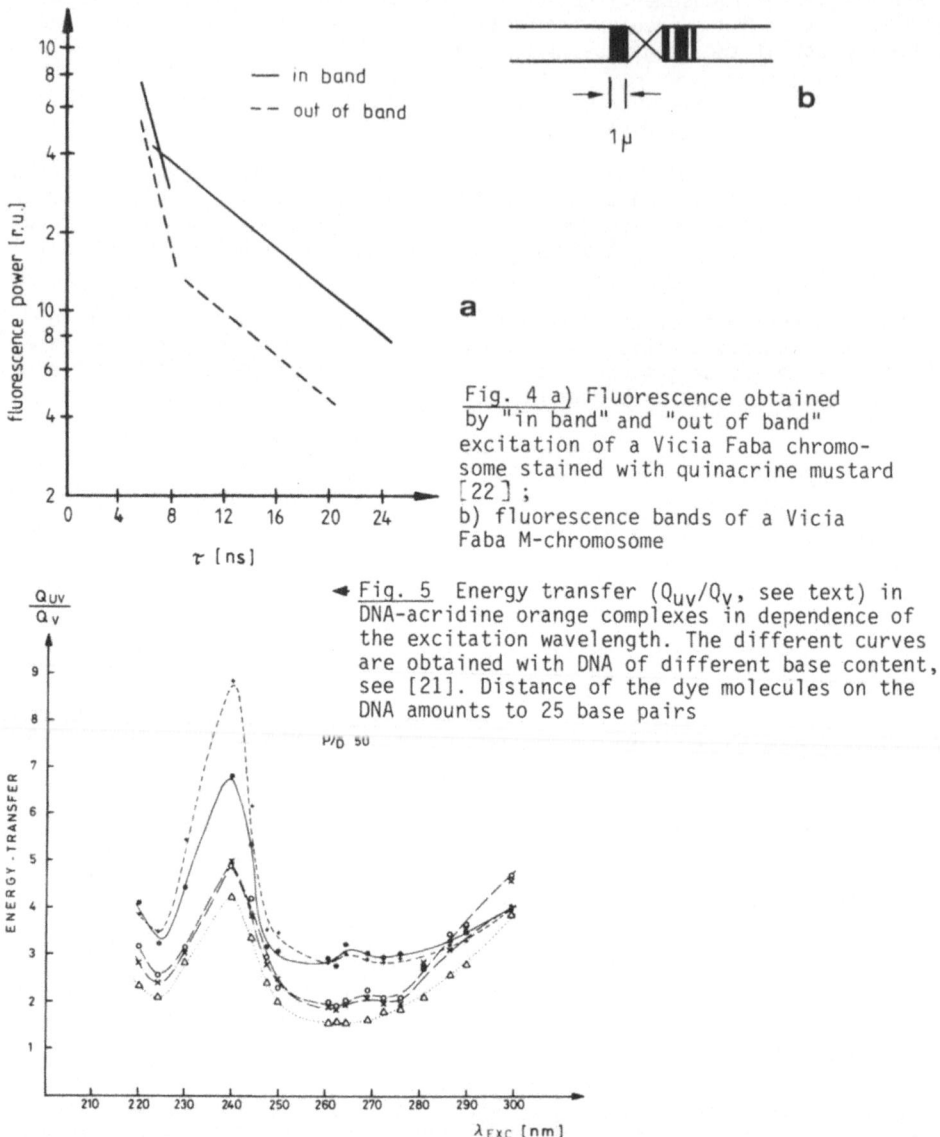

Fig. 4 a) Fluorescence obtained by "in band" and "out of band" excitation of a Vicia Faba chromosome stained with quinacrine mustard [22];
b) fluorescence bands of a Vicia Faba M-chromosome

Fig. 5 Energy transfer (Q_{uv}/Q_v, see text) in DNA-acridine orange complexes in dependence of the excitation wavelength. The different curves are obtained with DNA of different base content, see [21]. Distance of the dye molecules on the DNA amounts to 25 base pairs

The next point to be considered is energy transfer. Energy migration in nucleic acids plays an important role, e.g. for UV produced defects in DNA of cells and for the interaction between DNA and small molecules. There are two kinds of energy transfer, one from base to base along the DNA (exciton transfer) [11,21] and another from DNA to dye [20,21]. Both kinds of transfer depend on the excitation wavelength and on the DNA base sequence. The transfer from DNA to dye can be determined comparing the fluorescence quantum yields of the dye after excitation in the visible (Q_v) and in the (Q_{uv}) below 310 nm where the nucleic acids absorb. Fig. 5 gives an example of energy transfer in different DNA-acridine orange complexes.

162

One way to evaluate the exciton transfer is the comparison of the quantum yields of the dye fluorescence from complexes where the dye molecules are bound at different intervals along the DNA. The investigation of different DNA-acridine orange complexes using a frequency doubled dye laser as an excitation source (220-280 nm) renders an exciton transfer of about 80 base pairs along the DNA (50 % AT base pairs). This figure changes to 100 base pairs in AT-polynucleotides and to 60 pairs in GC polynucleotides. When the excitation takes place around 300 nm near the 0-0 energy of DNA a transfer of some hundred base pairs was observed [21].

Time resolved observations of energy transfer processes in DNA-acridine orange complexes were performed by SHAPIRO et al. [23]. They excited the complexes with 10 ps pulses from a mode-locked Nd:Yag laser at 265 nm. The measurement of the fluorescence risetime delivers the interval over which the exciton transfer along the DNA takes place.

Of special interest is the increase of the energy transfer from DNA to intercalated dye molecules and the increase of the transfer along the DNA near the 0-0 energy of DNA (see chapter 6 and [6]). The increase of both transfers is due to a decrease of radiationless processes in such big molecules when exciting at their 0-0-energy.

The dependence of the energy transfer on the base sequence allows the determination of the content of special double base pairs in different DNA [17]. Furthermore, energy-transfer measurements give information about molecular structures of the complexes, e.g., the position of the dye molecules in the plane between the bases (Fig. 6), when the transition dipole moments are considered [20].

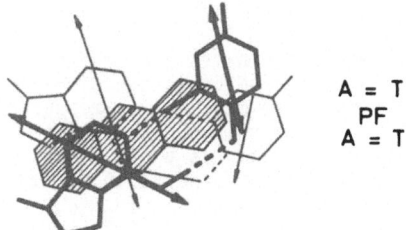

A = T
  PF
A = T

Fig. 6 Intercalation model.
Position of proflavine (hatched)
between an AT-AT base pair in
topview. The arrows represent
the transition dipole moments
[20]

A special kind of DNA-dye complexes are furocumarin complexes, which are used as drugs in photochemotherapy of dermatosis like psoriasis and vitiligo. The biological action of furocumarins occurs via their triplet state. SLOPER et al. [24] and BENSASSON et al. [25] determined the quantum yield of the triplet formation by laser flash spectroscopy.

4. Other Biomolecules

Photosynthesis and vision are two examples which are suitably studied by picosecond laser spectroscopy, as the lifetimes of the molecules involved are very short and the single processes occur very rapidly. Laser investigations of both processes are reviewed by PARSON [9], McCRAY et al. [1o] and CAMPILLO et al. [11].

In photosynthesis of higher plants light energy is absorbed and converted into chemical energy. The primary step includes a transfer of the

absorbed energy to the reaction centers of photosystem I and II. Information about these processes delivers the time dependence of the fluorescence
of chloroplasts. SEIBERT et al. [26] excited the fluorescence of isolated
spinach chloroplasts with a 4 ps pulse from a mode-locked Nd:Yag laser.

The time resolved fluorescence shows two maxima with a 90 ps delay between them. The first peak could be associated with components of photosystem
I having a lifetime of 10 ps, the second one with components of photosystem
II with a lifetime of 210 ps. The delay between the two peaks is related to
energy transfer between accessory pigments [26].

Photophysical processes that take place in chlorophyll solutions under
intense N_2-laser irradiation were investigated by HINDMAN et al. [27]. They
studied stimulated fluorescence and fluorescence quenching of chlorophyll a
and bacteriochlorophyll in solutions. Phenomena like line-width narrowing
and increase in fluorescence quantum yield with increasing pump power were
observed. Such processes are associated with high populations of molecules
in the first excited singlet state.

Further biomolecules which are suitably studied with laser excitation
are rhodopsin, bacteriorhodopsin, hemoglobin and myoglobin, for example.
They shall be only short mentioned because, until now, the most information
is not obtained from laser fluorescence studies, but from the observation
of optical absorption changes.

Excitation of myoglobin[29] and rhodopsin [30] was performed with picosecond pulses from a Nd^{3+}-glass laser at 530 nm, and the excitation of
bacteriorhodopsin [31] and hemoglobin [32] with single subpicopulses from a
cw mode-locked dye laser at 615 nm.

Rhodopsin is one of the light absorbing molecules in the process of vision. After light absorption a formation of several intermediate products
follows. The first steps of this process are expected to be in the picosecond range. The formation lifetime of the first intermediate was measured
to be less than 6 ps, for example [30] . Bacteriorhodopsin shows similari-
ties to rhodopsin and is of interest because of its role as a biological
energy converter. The light induced absorption at 615 nm appears in about
1 ps, for example [31] .

Kinetics of fast reactions were also studied in hemoglobin [32] and myoglobin [29] . In both molecules molecular oxygen is reversible bound to the
heme sites.

5. Selective Excitation

Selective laser action on biomolecules is an actual problem, because complex
biochemical reactions may be influenced in this way. Selective excitation
acts mainly in two ways: first, stimulation of single quantum states in one
kind of molecule and second, excitation of special kinds of molecules in a
mixture or of special parts in macromolecules.

Electronic absorption bands of biomolecules are normally broad and overlapping. Besides electronic excitation one can selectively excite the vibrational levels [4]. In Fig. 7 two ways of selective excitation are shown. In
the case of a two-step IR-UV excitation process the first step must be in
induced with ps pulses to keep the selectivity for the second step because
of the very fast vibrational relaxation processes.

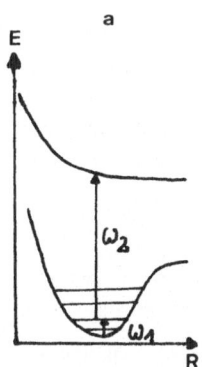
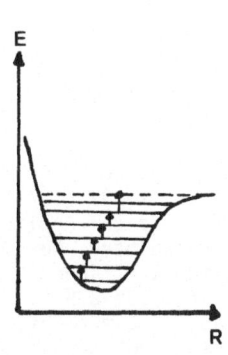

Fig. 7 Potential curves of complex molecules (E: energy, R: intermolecular distance.) a) Two-step, IR (ω_1) and UV (ω_2), excitation; b) multi-photon IR excitation [4]

One-step electronic excitation with frequency doubled dye lasers was performed in DNA-molecules[28]. The bases were excited at their first absorption maxima and at their 0-0 energies . Guanine can be excited without stimulating the other bases, because it has the lowest 0-0 energy. The selective excitation of special bases in DNA was tested by the interaction with dye molecules. The energy transfer in the three base pair units (see above) altered after favoured excitation of one kind of bases [28]. Multiquantum photoreactions in components of nucleic acids were reported by KRYUKOV et al. [33]. They used ps pulses from a mode-locked Nd:Yag laser (266 nm) and observed the photoproducts formed. Furthermore, a selective two-step dissociation was induced in dyes bound to polynucleotides either of AT or GC pair sequences [34].

6. Perspectives

Considering chromosome staining, additional fluorescence bands or a disappearance of bands have been observed in connection with special diseases, e.g., with the cancer disease Burkitt lymphoma (probably caused by a virus) where often a new band appears. Possible reasons are an alteration of the DNA base sequence or an increased dye binding in these chromosome regions. To elucidate these phenomena a selected laser excitation of chromosome parts seems promising.

The control of the replication process in DNA and an intended induction of mutations using laser excitation would be of great importance. The possibility of selective breaking of hydrogen bonds in DNA resulting in a splitting of the helix is discussed by LETOKHOV [4]. - Future techniques like an "optical mass spectrometer" and a "laser ion-microscope" are pointed out in Ref. 4, too.

In nucleic acids an excitation range of special interest lies around 300 nm [6] . This is the region where under natural conditions defects in DNA are induced from sunlight. The energy transfer in DNA and in some DNA-dye complexes increases in this part of the spectrum. In phototherapy and photochemotherapy irradiation in this wavelength range is often applied and therefore the interaction between nucleic acids and photosensitizers (e.g. furocoumarins [6, 25]) should be well known. The narrowband irradiation with lasers in photomedicine is under investigation [6] .

References

1. F.P. Schaefer (ed.): Dye Lasers (Springer, Berlin, Heidelberg, New York 1973)
2. H. Walther (ed.): Laser Spectroscopy of Atoms and Molecules (Springer, Berlin, Heidelberg, New York 1976)
3. A. Mooradian, T. Jaeger, P.Stokseth (eds.): Tunable Lasers and Applications (Springer, Berlin, Heidelberg, New York 1976)
4. V.S. Letokhov: Physics Today, No. 5. 23-32 (1977)
5. C.B. Moore (ed.); Chemical and Biochemical Applications of Lasers Vol. 1, 2, 4 (Springer, Berlin, Heidelberg, New York 1974, 1977, 1980)
6. A. Anders, P. Aufmuth, E.-M. Böttger, H. Tronnier: In this book
7. B.R. Ware: In ref. [5] , Vol. 2, pp. 199-239
8. A. Anders, A. Yasui, H. Zacharias, I. Lamprecht, W. Laskowski: In Radiation and Cellular Control Processes, ed. by J. Kiefer (Springer, Berlin, Heidelberg 1976) pp. 221-226
9. W.W. Parson: In ref. [5] , Vol. 1, pp. 339-372
1o. J.A. McCray, P.D. Smith: In Laser Application, Vol. 3, ed. by M. Ross (Academic Press, New York, San Francisco, London 1977) pp. 1-42
11. A.J. Campillo, S.L. Shapiro: In ref. [13], pp. 364-376
12. V. Wilke, W. Schmidt: Appl. Phys. 18, 177-181 (1979)
13. S.L. Shapiro (ed.): Ultrashort Light Pulses (Springer, Berlin, Heidelberg, New York 1977)
14. M.W. Berns: In Laser Applications in Medicine and Biology, Vol. 2, ed. by M.L. Wolbarsht (Plenum Press, New York, San Francisco, London 1974) pp. 1-4o
15. A. Andreoni, A. Longoni, C.A. Sacchi, O. Svelto, G. Bottiroli: In ref. [3], pp. 303-313
16. C.A. Sacchi, O. Svelto, G. Prenna: Histochem. J. 6, 251-258 (1974)
17. A. Anders: To be published
18. M. Ehrenberg, R. Rigler: In ref. [3]
19. A. Andreoni, C.A. Sacchi, O. Svelto: In ref. [5], Vol. 4
2o. A. Anders: Appl. Phys. 18, 333-338 (1979)
21. A. Anders: Opt. Commun. 26, 339-342 (1978)
22. A. Andreoni, C.A. Sacchi, S. Cova, G. Bottiroli, G. Prenna: In Lasers in Physical Chemistry and Biophysics, ed. by J. Joussot-Dubien (Elsvier Scientific Publ., Amsterdam, Oxford, New York 1975)
23. S.L. Shapiro, A.J. Campillo, V.H. Kollman, W.B. Goad: Opt. Commun. 15, 3o8-310 (1975)
24. R.W. Sloper, T.G. Truscott, E.J. Land: Photochem. Photobiol. 29, 1025-1029 (1979)
25. R. Bensasson, J. Land, A. Salet, SA E Melo, T.G. Truscott: In this book
26. M. Seibert, R.R. Alfano: Biophys. J. 14, 269-283 (1974)
27. J.C. Hindman, R. Kugel, A. Svirmickas, J.J. Katz: Chem. Phys. Lett. 53, 197-200 (1978)
28. A. Anders: Appl. Phys. 2o, 257-259 (1979)
29. W.G. Eisert, E.O. Degenkolb, L.J. Noe, P.M. Rentzepis: Biophys. J. 25, 455-464 (1979)
3o. P.M. Rentzepis: Biophys. J. 24, 272-284 (1978)
31. E.P. Ippen, C.V. Shank, A. Lewis, M.A. Marcus: Science 2oo, 1279-1281 (1978)
32. C.V. Shank, E. Ippen, R. Bersohn: Science 193, 50-54 (1976)
33. P.G. Kryukov, V.S. Lethokov, D.N. Nikogosyan, A.V. Borodavkin, E.I. Budowsky, H.A. Simukova: Chem. Phys. Lett. 61, 375-379 (1979)
34. A. Andreoni, G. Bottiroli, R. Cubeddu, S. de Silvestri, O. Svelto: In this book

Spectroscopic Properties of Quinacrine Mustard

A. Andreoni, R. Cubeddu, S. de Silvestri, and P. Laporta

Centro di Elettronica Quantistica e Strumentazione Elettronica
Istituto di Fisica del Politecnico
Milano, Italy

1. Introduction

Quinacrine Mustard (QM) is one of the most interesting dyes belonging to the Acridine family. In fact, QM binds specifically to the DNA and its fluorescence properties are dependent on the DNA base pair sequences [1]. In addition, QM is currently used as staining agent for metaphase chromosomes, giving the well-known fluorescence banding patterns [2]. Despite its importance, the spectroscopic properties of QM have not been completely investigated.

In this paper, we report results on fluorescence and absorption properties of QM in acetate buffer solution. The experimental data seem to indicate that three protonated species of this molecule are formed in the excited state.

2. Results and Discussion

The excitation spectra from 340 nm to 470 nm of QM (pH = 4.6, ionic strength 0.2 M) obtained by a conventional spectrofluorometer at two observation wavelengths (i.e. 480 nm and 540 nm) are shown in Fig. 1. The two spectra are

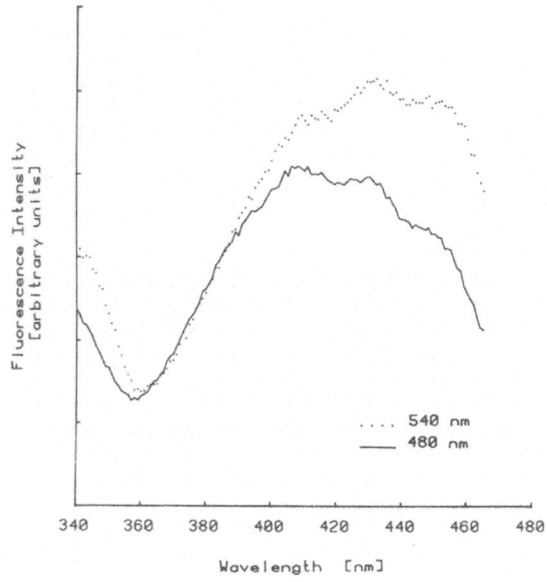

.... 540 nm
—— 480 nm

Fig.1 Excitation spectra of QM in 0.2 M acetate buffer solution (pH = 4.6) observed at 480 nm and 540 nm

167

markedly different. Both curves present three peaks, but with different relative intensities and shifted positions. The dependence of the excitation spectrum on the observation wavelength seems to indicate that the fluorescence is not emitted from a single excited electronic state. This conclusion, however, is in contrast with previous fluorescence decay measurements, which revealed a single exponential decay of the spectrum-integrated fluorescence under excitation at 419 nm [3]. The time-resolved fluorescence spectrum at different excitation wavelengths provided valuable information for understanding of the emission mechanism. The fluorescence decay was detected at several observation wavelengths spanning the whole fluorescence spectrum (i.e. from 460 nm to 725 nm). The measurements were performed under excitation at 383 nm, 420 nm and 460 nm.

The excitation pulses of $\sim$ 150 ps duration, $\sim$ 40 kW peak power and repetition rate up to 100 Hz, were provided by a dye laser specially designed by us [4]. The dye laser was pumped by an atmospheric pressure nitrogen laser, which generates pulses of 500 ps duration [5]. The fluorescence signal observed at 90° through a monochromator was detected by a Varian 154 M photomultiplier (400 ps FWHM). The photomultiplier output was sent to a home-made digital signal averager with 100 ps resolution [1].

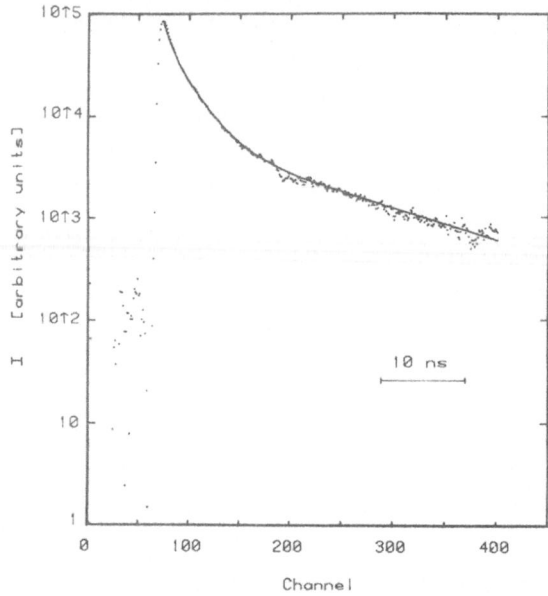

Fig.2 Logarithmic plot of the fluorescence decay under excitation at 420 nm observed at 480 nm. Experimental data (dots); three-exponential-component fitting (unbroken line)

The fluorescence decay was found to be independent of both the QM concentration (in the 10^{-5} - 10^{-3} M range) and the excitation intensity. The peak values of the signals were linear with the excitation peak power. Under excitation at 420 nm the fluorescence decay can be fitted, in the 460 - 520 nm observation range, by three exponential components with time constants τ_1 = 0.8 ± 0.2 ns, τ_2 = 2.7 ± 0.2 ns and τ_3 = 16.4 ± 0.2 ns. Fig. 2 shows a logarithmic plot of the experimental fluorescence curve observed at 480 nm and the computed three-component fitting. The values of the relative amplitudes A_1, A_2 and A_3 of the three components vary in this interval as reported in Fig. 3. At longer observation wavelengths (i.e. from 520 nm to 725 nm) only the intermediate and the slow components are present in the fluorescence decay. For 383 nm excitation, the fluorescence curves are still well fitted by

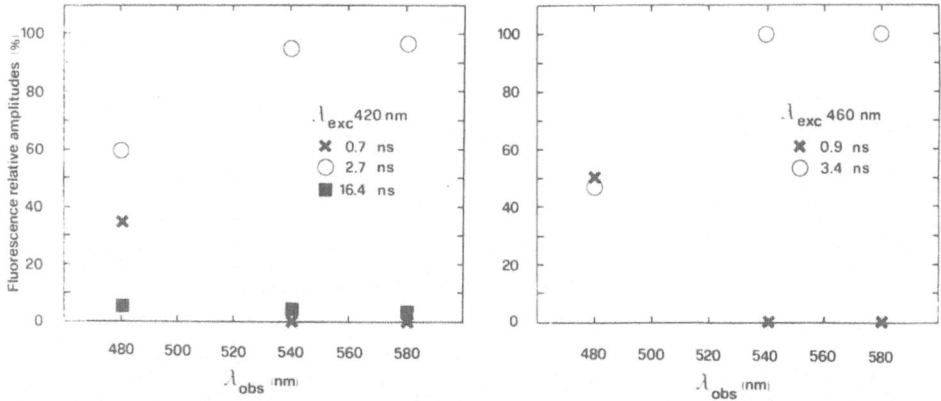

Fig.3 Relative fluorescence amplitudes of the three exponential components at different observation wavelengths for excitation at 420 nm and 460 nm

three exponential components, with very similar time constants, but different amplitudes as compared to the case of 420 nm excitation. For 460 nm excitation the slowest decay was never observed over the whole fluorescence spectrum. The relative amplitudes A_1 and A_2 of the two components that are still present vary in the observation range from 480 nm to 580 nm as shown in Fig. 3. All these results are summarized in Table 1.

Table 1 Fluorescence parameters of Quinacrine Mustard at different excitation and observation wavelengths

λ_{exc}[nm]	λ_{obs}[nm]	τ_1[ns]	A_1(%)	τ_2[ns]	A_2(%)	τ_3[ns]	A_3(%)
383	480	0.8	52.5	2.7	38.2	16.4	9.3
	540			2.9	90.9	16	9.1
420	480	0.7	34.1	2.6	59.3	16.4	6.6
	540			2.8	94	16.1	6
	580			3.2	94.9	14.8	5.1
460	480	0.9	50.5	3.3	49.5		
	540			3.4	100		
	580			3.6	100		

Analysis of the experimental data seems to indicate the presence of three molecular species with different fluorescence spectra and decay times. Chromatographic analysis revealed however no impurities at a level that could account for our results. We are then led to assume that the three species are different forms of the same molecule. In this regard, we should like to point out that: (i) The three nitrogen atoms (Fig.4) in the neutral molecule can coordinate three protons leading to the formation of monocationic, dicationic and tricationic forms. (ii) Protonated species of the similar dye, Quinacrine, are known to exist in ground state [6]; their relative concentrations depend on the pH value as evidenced by the changes in the absorption spectrum [7].

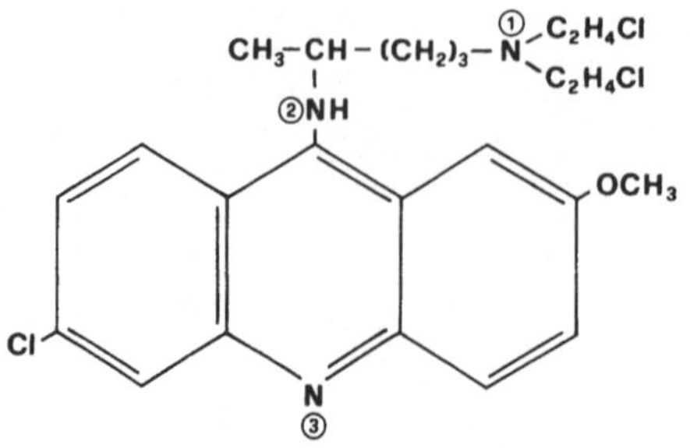

Fig.4 Chemical structure of QM neutral molecule. The three protonation sites are indicated with the circled number: ① monocation; ① ② dication; ① ② ③ trication

(iii) Multiexponential fluorescence decay has been recently reported for Quinacrine [8]. It must be noted that QM, as well as Quinacrine [7], exists in only one protonated species in the ground state at pH = 4.6 . According to ref. [7], this species has been attributed to the dicationic form of the molecule. We therefore suggest that the different protonated species are formed by multiproton transfer in the excited state. Indeed, the pK values for the formation of ionic species may be quite different in the excited state, as compared with those in the ground state [9]. Finally we should like to point out that, if our assumption is correct, the protonation process must occur on a time-scale at least comparable with that of the intraband relaxation. In fact, only in this case is it possible to expect a yield of formation of the different species that depends on the excitation wavelength [see Table 1].

Acknowledgement

We should like to thank Prof. O.Svelto for stimulating discussions and for reading the manuscript.

References

[1] A. Andreoni, C.A. Sacchi and O. Svelto, in: Chemical and Biochemical Applications of Lasers, Vol. 4, ed. C. Bradley Moore (Academic Press Inc.,New York, 1979) p. 1

[2] T. Caspersson, L. Zech, E.J. Modest, G.E. Foley, U. Wagh and E. Simonsson, Exptl. Cell Res. 58, 141 (1969)

[3] A. Andreoni, S. Cova, G. Bottiroli and G. Prenna, Photochem. Photobiol. 29, 951 (1979)

[4] R. Cubeddu, S. De Silvestri and O. Svelto (to be published)

[5] R. Cubeddu, S. De Silvestri, Opt. Quant. Electr. 11, 276 (1979)

[6] A. Albert: The Acridines (E. Arnold Ltd., London, 1966) p. 161

[7] A.C. Capomacchia and S.G. Schulman, Anal. Chim. Acta 77, 79 (1975)

[8] D.J. Arndt-Jovin, S.A. Latt, G. Striker and T.M. Jovin, J. of Histochem. and Cytochem. 27, 87 (1979)

[9] A.J. Campillo, J.H. Clark, S.L. Shapiro and K.R. Winn, in: Picosecond Phenomena, eds. C.V. Shank, E.P. Ippen and S.L. Shapiro (Springer-Verlag, Berlin Heidelberg New York, 1978) p. 319

Optical-Microwave Double Resonance Spectroscopy of in Vivo Chlorophyll*

R.H. Clarke, S.P. Jagannathan, and W.R. Leenstra

Department of Chemistry, Boston University
Boston, MA 02215, USA

1. Introduction

The utilization of high resolution spectroscopic techniques in the investigation of photosynthetic systems has become an increasingly useful approach for the understanding of the photophysics of photosynthesis [1]. Laser techniques have afforded both narrow band excitation and short-time phenomena to be investigated in photosynthesis, providing a range of information on structural and dynamical features of the photosynthetic apparatus in vivo [1].

Since the original observations of photoexcited triplet state EPR signals in purple photosynthetic bacteria [2] [3] [4], a considerable research effort has focused on use of triplet state properties to determine structural and dynamical features of chlorophylls in vivo [5]. In such experiments the chlorophyll triplet state becomes, effectively, an in situ paramagnetic probe into the surroundings and interactions within the photosynthetic systems.

In our laboratory we have initiated and developed a combination of laser and microwave spectroscopy, optically-detected magnetic resonance spectroscopy (ODMR), for the study of the chlorophyll triplet state. We have utilized this approach to investigate structural features of chlorophyll in vitro, both as isolated [6] [7] and aggregated [8] systems, and chlorophylls in vivo [9] [10], providing the first ODMR spectrum of the photoexcited triplet state of an in vivo photosynthetic system [11]. The use of laser-microwave double resonance allows a high degree of resolution and sensitivity in obtaining chlorophyll triplet state spectra in the various media.

Although there has been much activity in the application of triplet state spectroscopy (both EPR and ODMR) to the chlorophyll molecule, chlorophyll aggregates and to photosynthetic bacteria [5], the investigation of triplet state chlorophyll in green plant systems has been far more limited, due, in part, to the higher complexity of the photosynthetic apparatus involved. Hoff and van der Waals observed a chlorophyll a ODMR signal in reduced chloroplast preparations [12], interpreted as arising from photosystem II; Schaafsma and coworkers have reported zero-field ODMR transitions in algae preparations [13] and Hoff et al have investigated by ODMR triplet state signals from chloroplast preparations, observing signals which are interpreted as non-chlorophyll components [14]. All data to date is in preliminary form, and a complete sur-

*Supported in part by the U.S.Department of Energy under contract number EY-76-S-02-2570 and by the donors of the Petroleum Research Fund, administered by the American Chemical Society. Laser and ODMR spectroscopic equipment developed under grants from the U.S. Army Research Office.

171

vey of triplet state signals, detected on the various fluorescence peaks of green plant systems has not been fully explored.

In the present paper we present initial results of an attempt to study systematically by laser-microwave double resonance techniques, the range of low temperature fluorescence bands commonly reported for green plant preparations. These include ODMR spectra of fluorescence considered to arise from photosystem I (PSI), photosystem II (PSII) and the light harvesting chlorophyll protein (LHCP) complex [15]. Both triplet state ODMR transition frequencies and individual triplet spin sublevel intersystem crossing rates have been obtained for chloroplast and subchloroplast preparations. These results are discussed with reference to possible structural features of in vivo chlorophylls.

2. Experimental Methods

The experimental arrangement for laser-microwave double resonance, using a pair of argon ion lasers has been described previously [6] [8].

Preparation of chloroplast and subchloroplast samples followed published procedures. Chloroplasts were prepared by homogenization of spinach leaves suspended in a sucrose-tricine- NaCl solution, followed by centrifugation [16]. Chemical reduction of the chloroplast preparations for the chloroplast ODMR experiments was accomplished by addition of sodium dithionite.

Photosystem I enriched particles were prepared by the method of Bengis and Nelson [17]. Chloroplasts were treated with Triton X-100, followed by chromatography on a DEAE-cellulose column. The most active fraction was centrifuged on a sucrose gradient.

3. Results and Discussion

The ODMR transition frequencies observed on the various fluorescence peaks for reduced chloroplasts and PS I enriched particles at 2 K are presented in Table 1.

Several points are evident from the data in Table 1. The first is that, despite the use of column separation techniques in preparations such as that described by Bengis and Nelson [17], the preparations contain appreciable amounts of solubilized chlorophylls [14]. These chlorophylls have an enhanced fluorescence intensity compared to in vivo chlorophylls and dominate the ODMR spectra observed on the fluorescence. The fluorescence band detected at 650 nm clearly originates from chlorophyll b molecules, both on the basis of its fluorescence wavelength [18] and from the ODMR frequencies detected at that wavelength. The chlorophyll b zero-field splitting is larger than that for chlorophyll a [7] and provides ODMR transitions at zero field to higher frequency than chlorophyll a, with typical values as observed in Table 1. Further, in Table 2, the individual spin sublevel T_1 - S_0 rate constants are presented for the triplet species detected at 650 nm. Again, these rates are in the range observed for chlorophyll b in hydrocarbon solution [7]. Both the ODMR transition frequencies and intersystem crossing rate data associated with the 650 nm peak confirm its assignment as free chlorophyll b pigment.

The remaining fluorescence bands reported in Table 1 all give ODMR transition frequencies in the range expected for the chlorophyll a molecule [7]. The

Table 1 ODMR transitions observed in chloroplast and subchloroplast particles at 2 K

System	excitation wavelength (nm)	detection wavelength (nm)	ν_1 (Mhz)	ν_2 (Mhz)
PS I enriched particles	457.9	650	854	1063
	514.5	670	766	1013
	514.5	725	767	1008
	454.5	670	769	1010
	454.5	680	720	983
Reduced chloroplasts	514.5	684	no ODMR observed	
	514.5	723	726	957

Table 2. T_1 - S_0 intersystem crossing rate constants determined from ODMR transitions at 2 K

system	detection wavelength (nm)	$k_x (sec^{-1})$	$k_y (sec^{-1})$	$k_z (sec^{-1})$
PS I enriched particles	650	550 ± 110	730 ± 150	45 ± 10
"	670	600 ± 120	650 ± 150	60 ± 20

shortest wavelength band, 670 nm, dominates the fluorescence spectrum of the particles prepared by the methods described in ref. [17]. It is reasonable to assign it, on the basis of its intensity and ODMR frequencies, to solubilized chlorophyll a pigments. Indeed, the ODMR transitions observed on the 725 nm fluorescence shoulder are, within experimental error, the same as observed at 670 nm, and, most likely arise from the same species, with the 725 nm intensity corresponding to a vibronic band built on the 670 nm peak, rather than an independent species in the sample.

The only ODMR transitions detected in our experiments at other than the frequencies seen at 670 nm occur at 680 nm in the PS I enriched samples and at 723 nm in reduced chloroplast preparations. Using the assignments of Butler et al [15], one expects the 680 nm (low temperature) fluorescence band to be associated with the LHCP complex and the 723 nm band to arise from within PS I. Both peaks provide ODMR transition frequencies lower than those observed at 670 nm and are clearly different in vivo species, most likely from antenna systems. Presumably, the PS I signals are observed in chloroplasts and not in the PS I enriched preparations, because in the latter case the true PS I fluorescence is masked by the strong emission from the solubilized pigments.

We have measured the spin sublevel decay rates for the triplet states associated with the 670 nm fluorescence band to determine if the pigments involved were free chlorophyll a. The results are shown in Table 2. The rates observed at 670 nm are consistently smaller than is observed for chlorophyll a in hydrocarbon solvents by several laboratories (k_x=661, k_y=1255, k_z=241 [7]).

In fact, the values are remarkably similar to those observed for the ODMR spectra detected at 650 nm, assigned as chlorophyll b. Since the chlorophyll b rates show that, under the sample conditions, the usual molecular rates can be obtained, one must conclude that the makeup of the solubilized species fluorescing at 670 nm is not simply free chlorophyll a molecule.

The similarity in dynamical features observed at 650 and 670 nm leads us to consider the possibility that the 670 nm fluorescence originates in fractionated particles of the LHCP complex. Such particles would contain both chlorophyll a and b, interacting within a protein environment. Since the chlorophyll a triplet state is known to be lower in energy than that for chlorophyll b [19], the triplet excitation within the 670 nm species may be trapped at chlorophyll a sites, providing chlorophyll a ODMR frequencies, but may be populating the chlorophyll a traps via energy transfer from nearby chlorophyll b pigments. Thus, the dynamics observed would reflect the longer-lived chlorophyll b species as it transferred its triplet energy, rather than the purely intramolecular $T_1 - S_0$ relaxation of the shorter-lived chlorophyll a. This explanation, if appropriate, would place important restrictions on the relative positioning of the chlorophyll a and b pigment molecules within the LHCP, since triplet energy transfer is well known to be a very short range phenomenon [20].

4. References

1. See for recent references CIBA Foundation Symposium 61 (Excerpta Medica, Amsterdam, 1979)
2. P.L. Dutton, J.S. Leigh, and M. Siebert, Biochem. Biophys. Res. Commun., 46, 406 (1972)
3. P.L Dutton, J.S. Leigh, and D.W. Reed, Biochim. Biophys. Acta, 292, 654 (1973)
4. J.S.Leigh and P.L. Dutton, Biochim. Biophys. Acta, 357, 67 (1974)
5. H.Levanon and J. R. Norris, Chem. Rev., 78, 185 (1978)
6. R.H. Clarke and R.H. Hofeldt, J. Chem. Phys., 61, 4582 (1974)
7. R.H. Clarke, R.E. Connors, T.J. Schaafsma, J.F. Kleibeuker, and R.J. Platenkamp, J. Am. Chem. Soc., 98, 3674 (1976)
8. R.H. Clarke, D.R.Hobart, and W.R. Leenstra, J. Am. Chem. Soc., 101, 2416 (1979)
9. R.H. Clarke and R.E. Connors, Chem. Phys. Lett., 42, 69 (1976)
10. R.H. Clarke and D.R. Hobart, FEBS Lett., 82, 155 (1977)
11. R.H. Clarke, R.E. Connors, J.R. Norris, and M.C.Thurnauer, J. Am. Chem. Soc., 97, 7178 (1975)
12. A.J. Hoff and J.H. van der Waals, Biochim. Biophys. Acta, 423, 615, (1976)
13. S.J. van der Bent, T.J. Schaafsma, and J.C. Goedheer, Biochem. Biophys. Res. Commun., 71, 1147 (1976)
14. A.J. Hoff, Govindjee, and J.C. Romijn, FEBS Lett. 73, 191 (1977)
15. W.L.Butler and M. Kitajima, Proc. Third Intl. Congr. Photosynthesis, M. Avron, ed. (Elsevier, Amsterdam, 1975), pp. 13-15
16. T.Yamashita and W.L Butler, Plant Phys., 43, 2037 (1968)
17. C. Bengis and N. Nelson, J. Biol. Chem., 253, 2783 (1975)
18. J. Brown, Photochem. Photobiol., 26, 519 (1977)
19. A.A. Krasnovskii, N.N. Lebedev, and F.F. Litvin, Dokl. Akad. Nauk SSSR, 216, 1406 (1974)
20. J. Jortner, S.A. Rice, J.L. Katz, and S.I. Choi, J. Chem. Phys., 42, 309 (1965)

Evaluation of the Chromatin Functional State by Means of Fluorescence Decay Time Analysis

G. Bottiroli[1], P.G. Cionini[2], F. Docchio[3], and C.A. Sacchi[3]

[1] Centro die Studio per l'Istochimica del C.N.R., Instituto di Anatomia Comparata, Universitā die Pavia

[2] Instituto die Genetica, Universitā die Pisa

[3] Centro die Studio per l'Elettronica Quantistica e Strumentazione Elettronica del C.N.R., Instituto di Fisica, Politecnico di Milano

1. Introduction

This paper discusses the application of laser microfluorometric techniques, together with the use of fluorescent probes, to investigate some functional aspects of biological macromolecules at the cellular level.

Following EDELMAN and McCLURE [1], fluorescent probes are defined as "small molecules which undergo changes in one or more of their fluorescent properties as a result of noncovalent interaction with protein or other macromolecule". This characteristic is determined by the lifetime of the excited state, which is sufficiently long for a variety of physical and chemical interactions to take place before the emission [2].

For this reason, fluorescent probes have become of great interest in the biological field, since they can provide evaluations of the complex biological material that go beyond the mere dosage. These probes have been widely employed in "in vitro" studies of well-defined solutions, whereas their extension to more complicated systems, like intact cells, has not advanced so far [3].

In this case, the difficulty of finding unambiguous relationships between fluorescence parameters and the biological structure arises mainly from the complexity and the dishomogeneities of the latter.

For this reason, conventional histochemical techniques, mainly based on fluorescence intensity measurements, do not make it possible to say whether the intensity depends on a different bond stoichiometry, or on a different quantum yield of the bound dye, as a consequence of different binding mechanisms. On the other hand, a detailed analysis of fluorescence emission based on the measurement of the decay time has been more fruitful for a general interpretation of the results, and particularly useful for "in situ" studies [4,5]. Among the possible applications of the fluorescence decay time analysis, we have been concentrating on the "in situ" evaluation of the structural and functional characteristics of chromatin by using fluorescent probes, as already suggested in ref. [6].

Chromatin is a cellular component that is distributed throughout the nucleus of the cell. It consists of a continuous linear DNA double strand and of associations of histone proteins, non-histone proteins, RNA, and enzymes such as DNA- and RNA-polymerase. Histone proteins are present in equal pro-

*Work supported by the Italian National Research Council through the Special Program "Tecnologie Biomediche".

175

portions, by weight, to DNA; the other components are present in variable amounts with respect to DNA. This is particularly the case with non-histone (or acidic) proteins, which are of great importance for the template activity of the chromatin DNA, a process that, by means of the transcription of RNA, triggers the synthesis of proteins. Non-histone proteins, in fact, have regulating functions, and are present at a considerably higher content in the active form of chromatin [7,8] .

Chromatin fractions in different functional states show different chemico-physical and morphological characteristics. Among these, the packing ratio, which is lower in active and higher in inactive chromatin, plays an important role, because it leads to the fact that the melting temperature of transcribed chromatin is relatively low, and that transcribed chromatin is readily attacked by nucleases such as DNase I and DNase II. In all these respects, the properties of isolated transcribed chromatin approach those of free DNA. At the cellular level, active chromatin shows a looser and less packed structure than inactive chromatin [9,10,11] .

There is evidence, from all these differences, that a chromatin fraction presents a structure that is related to its functional state.

As an approach to the study of the functional state of chromatin by means of decay time analysis, Quinacrine Mustard (QM), a well known intercalating and alkylating acridinic derivative, has been used. In fact, in recent years, research has been performed in our laboratories on the behaviour of QM, both on extracted and purified DNA and on particularly simple biological structures like bacteria [12,13] . So far, the following results have been obtained:
i)   QM specifically interacts,under suitable conditions, with DNA.
ii) The decay time of the total fluorescent emission depends on the base composition of the bound DNA.
iii) The decay time of the DNA-QM complex depends, the base composition being equal, on the stoichiometric ratio r between the number of QM intercalated molecules and the number of DNA base pairs, and therefore on the degree of intercalation of DNA.

An energy transfer mechanism has been proposed as a possible explanation for the dependence of the decay time of the complex on the two aforementioned parameters.

Nevertheless, apart from the interpretation of the intercalation-fluorescence time relationship, the analysis of the decay time of the QM-DNA complex could be used as a useful method for the evaluation of the chromatin activity in biological samples.

In fact, different degrees of activity, which correspond to different structures, should also affect the chromatin's stainability and, therefore, the fluorescence decay time.

As a first step towards testing this assumption, it is important to eliminate the dependence of the decay time on the other parameter mentioned, i.e. the base composition of the DNA of the biological sample. For this purpose, polytene chromosomes obtained with a squash of giant embryo suspensor cells of Phaseolus coccineus (Fig. 1) were investigated. In this material, differences in the degree of activity easily recognizable on the basis of morphological characteristics, may be present not only in one and the same chromosome region in different cells, but also in the corresponding segments of the two homologous in one and the same cell [14] .

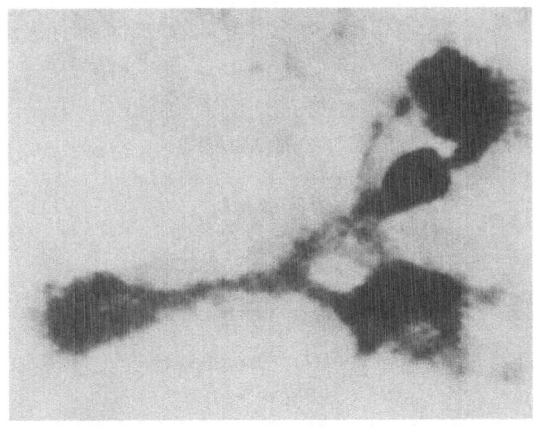

Fig. 1 Polytene chromosome of the embryo suspensor cell of Phaseolus coccineus along which differences in the degree of activity are morphologically recognizable. Feulgen reaction, × 1150.

In this case, since the base composition of the DNA is the same, a variation in the fluorescence decay time, if it exist, should only be due to the chromatin structure of the chromosome segments.

2. Experimental

The chromosomes to be investigated were obtained from Phaseolus coccineus developing seeds fixed for 1 h in ethyl alcohol: acetic acid 3:1 (v/v) solution. Squashes were prepared by dissecting the embryo suspensor and macerating the material with 5% pectinase solution at pH 4.0 and 40°C for 1 h.

The samples were subsequently stained with a 5×10^{-5}M solution of QM in acetate buffer at pH 4.6, and then washed in the same buffer solution. A 0.2 M ionic strength was used to avoid interferences due to non-intercalating bonds, such as bonds external to DNA and RNA [13] . The measurements were made on the nucleolus organizing region of S_1 chromosomes.

Analysis of the decay times of the fluorescent emission was performed by means of the pulsed laser microfluorometer shown in Fig. 2 [4], which is briefly described here. The output beam of a subnanosecond pulsed dye laser, pumped by a N_2 laser, is sent to the side entrance-window of a Leitz MPV microscope equipped with a vertical illuminator for fluorescence excitation in incident light (dichroic mirror TK450). The wavelength of the beam is 419 nm, corresponding to the peak of the QM absorption spectrum. The pulses have a duration of 0.5 nsec FWHM, and a repetition rate of 25 pps. The beam is focused by the microscope objective down to a spot of 0.5 μm^2 on the object plane. The fluorescent emission of the sample is collected by a fast RCA 70045D photomultiplier (risetime ~ 0.3 ns) mounted at the top of the Leitz MPV system.

A digital averager developed in our laboratory [15] is used to average a suitable number of waveforms, in order to reduce the relative fluctuations of the signals due to the low number of photons emitted per pulse. The fluorescence waveforms are sampled at a suitable number of points by means of a sampling oscilloscope. Analog-to-digital conversion of the sampled values is performed by a gated voltage-to-frequency converter. The sum, or average, of the corresponding values is made by means of a multichannel analyzer. The digitized and averaged waveforms are stored in magnetic tapes, and the data are eventually transferred to an off-line computer for additional processing and graphics.

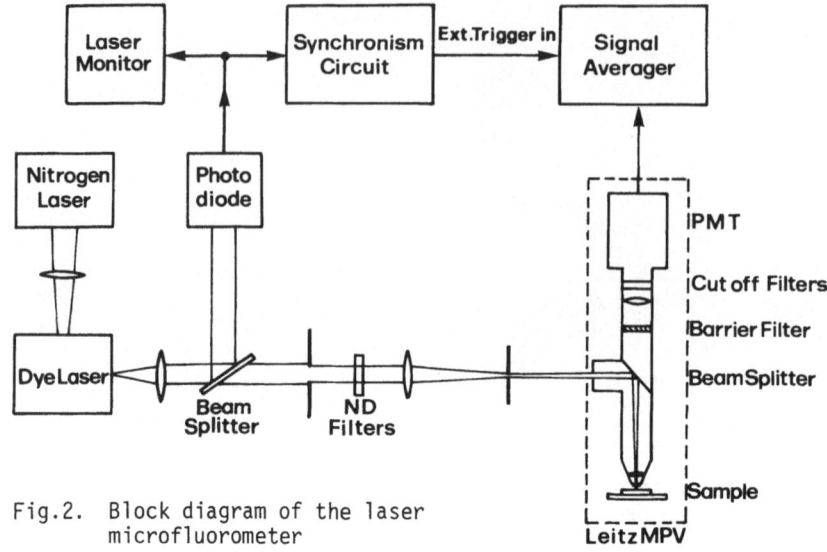

Fig.2. Block diagram of the laser microfluorometer

3. Results and Discussion

Two typical fluorescence waveforms obtained with our apparatus from the corresponding segments of homologous chromosomes with active and inactive chromatin are shown in Fig. 3. The decay time for active chromatin (○) turned out to be 12.6 + 0.2 ns, while for inactive chromatin (●) was 9.7 + 0.2 ns. (i.e. substantially shorter than the previous one).

According to the results obtained "in vitro" [13] , these results suggest a lower stainability (smaller r value) of the active chromatin, despite its looser structure. This can be explained if we consider that, as previously mentioned, active chromatin has a larger amount of non-histone (i.e. acidic) proteins. It is known that these proteins are more tightly bound to DNA than are histones, and that the steric situation of the DNA-non-histone-protein complex is different from the DNA-histone-protein complex [9,16] . These factors could therefore determine a lower availability to intercalation of DNA for active chromatin than for inactive chromatin, contrary to what a simple morphological analysis would suggest.

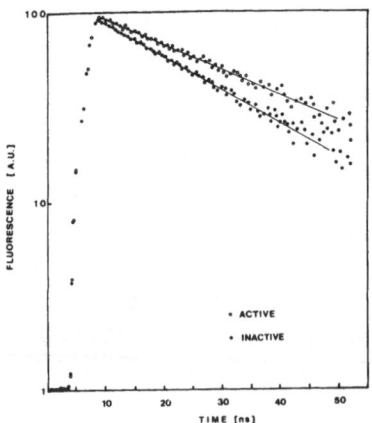

Fig. 3. Fluorescence decay times of QM in active and inactive chromatin from chromosomes of the type represented in Fig. 1.

178

The use of a suitable model, like the chromosome of Phaseolus coccineus, has thus indicated that the fluorescence decay time analysis of a DNA-fluorescent-probe complex is a powerful technique for the investigation of the dynamic aspects of biological structures. This technique, if extensively verified, could open up important prospects when applied to nuclei "in toto": i.e., those of enabling the cell cycle phase to be identified when no variations in the amount of DNA are involved.

In this case it could be possible to distinguish by means of simple and short procedures, between quiescent and cycling cells. This, in turn, has a far-reaching implication when the presence of cycling cells, or any abnormal increase in their amount, can be related to the existence of evolution of pathological conditions, such as, for instance, leukemia [17] .

A further possible application of the method is related to the problem of optimizing the action mechanism of several antiblastic drugs, whose fluorescence properties are similar (e.g. anthracyclines) to acridines [18] . This could be achieved with a more detailed knowledge of the extent to which the active drug amount depends on the functional state of the substrate.

REFERENCES

1. Edelman,G.M. and McClure, W.O., Accounts Chem. Res. 1, 65-70 (1968)
2. Brand, L. and Golke, J.R., Ann. Rev. Biochem. 41, 843-868 (1972)
3. Le Pecq, J.B., in: "Biochemical Fluorescence" vol. 2, ed. Chen, R.F. (M.Dekker, New York, Basel 1976), p. 711
4. Sacchi, C.A., Svelto, O. and Prenna G., Histochem. J. 6, 251-258 (1974)
5. Andreoni, A., Sacchi, C.A. and Svelto, O., in: "Chemical and Biochemical Applications of Lasers", vol. 4, ed. Bradley Moore, C. (Academic Press, New York, 1979), p. 1
6. Andreoni, A., Sacchi, C.A., Svelto, O., Bottiroli, G. and Prenna, G., Sov. J. Quantum Electr. 8 (10), 1255-1259 (1978)
7. Johnson, J.D., Douvas, A.S. and Bonner, J., Int. Rev. Cytol., suppl. 4, 273-361 (1974)
8. Stein, G.S., Spelsberg, T.C. and Kleinsmith, L.J., Science 183, 817-824 (1974)
9. Revees, R. and Jones, A., Nature 260, 495-500 (1976)
10. Bonner, J., in: "Chromatin Structure and Function" ed. Nicolini, C.A. (Plenum Press, New York, London, 1979), p. 15
11. McKnight, S.L., Martin, K.A., Beyer, A.L. and Miller ir., O.L., in: "The Cell Nucleus", vol. 7, ed. Busch, H. (Academic Press, New York, San Francisco, London, 1979), p. 97
12. Bottiroli, G., Prenna, G., Andreoni, A., Sacchi, C.A. and Svelto, O., Photochem. Photobiol. 29, 23-28 (1979)
13. Andreoni, A., Cova, S., Bottiroli, G. and Prenna, G.; Photochem. Photobiol. 29, 951-957 (1979)
14. Durante, M., Cionini, P.G., Avanzi, S., Cremonini, R. and D'Amato, F., Chromosoma 60, 269-282 (1977)
15. Andreoni, A., Longoni, A., Sacchi, C.A. and Svelto, O., in: "Laser Biomedicine Engineering" ed. by L.Goldman (Springer-Verlag, Berlin, 1980)
16. Chiu, J.F. and Hnilica, L.S., in: "Chromatin and Chromosome Structure", eds. Li, H.J. and Eckhardt, R. (Academic Press, New York, San Francisco, London, 1977), p. 193
17. Baserga, R., "Multiplication and Division in Mammalian Cells" (M.Dekker Inc., New York, Basel, 1976)
18. Comings, D.E. and Drets, M.E., Chromosoma 56, 199-211 (1976)

Part VI

Raman and Picosecond Spectroscopy

Resonance-Raman Spectroscopy in Biological Problems: Use of Labels

G. Zerbi[1]

Instituto di Chimica, Universitã Trieste, Italy

G. Masetti and L. Nannicini

LAFBIC, CNR, Via Buonarroti 9, Pisa, Italy

G. Dellepiane

CNR, c/o Chimica Industriale, Politecnico, Milano, Italy

1. Introduction

In the study of the structure-function relationships of biological materials most of the physical measurements provide overall information which must then be interpreted on the basis of suitably chosen models. This is particularly true in the case of biological substances in solution where all symmetry relationships are destroyed thus denying any possibility of carrying out coherent scattering measurements of both X-ray or neutrons. While the vibrational spectrum (Infrared and Raman) of biological molecules has been widely used in the past twenty years as a structural tool, the interpretation of the experimental data in terms of microstructure often fails because of the intrinsic complexity of the systems. The recent introduction of laser Raman spectroscopy in this field has given a new momentum to this kind of studies because it has opened the way to focus the attention in great detail at a specific site of the molecule.

The basic fact is that with the use of tunable lasers it becomes possible to select an exciting wavelength in resonance with one of the electronic transitions which mainly takes place in a particular site of the macromolecule. Having reached a resonance condition, the vibration modes which are coupled with that particular electronic transition become strongly Raman enhanced [1], while all other modes are missing. The application of this phenomenon allows us to study biological systems in physiological conditions.

The main limitation of this kind of approach to structural studies of biological materials is due to the present very scarce availability of laser sources in the UV region where most of the electronic transitions of the intrinsic chromophores take place. One way to overcome this difficulty is to label the material with suitable chemical groups [2] (external chromophores) which absorb at higher wavelength, thus allowing for Resonance Raman Spectroscopy the use of the commonly available laser sources. The comparison of the Resonance Raman Spectrum of the free label and of the biological material suitably labelled in sites of critical interactions provides the way to obtain information either on the site itself or on the topology of the molecular environment around the label.

We wish to report here part of the results of our studies by Resonance Raman Spectroscopy on the enzyme aldolase which has been labelled with various benzeneazo groups (Azoaldolase). Aldolase is the enzyme which catalyzes in the

---

[1] To whom correspondence should be adressed: Instituto di Chimica Industriale del Politecnico, P.zza Leonardo da Vinci, 32, Milano, Italy

muscles the key reaction of reversible cleavage of fructose-1,6 bis-phosphate. The primary structure of the rabbit muscle FruP$_2$ aldolase is known; it has molecular weight 160.000 and it is composed of four subunits. Each subunit is made up by 361 residues whose known sequence contains 8 cysteine residues and 11 histidine residues [3]. FELICIOLI et al. [4] have recently shown that the introduction of azo labels is possible in aldolase and that the modified material still keeps its enzymatic activity [4]. The same group suggests that the most stable azoaldolase is that which contains two azobenzene groups attached to cysteine 237 and 287 [4]. It has also been shown that azoaldolase is photosensitive and can be photochemically destroyed according to the pH, the dose of irradiation and its wavelength [5].

The purpose of our work in this field is to contribute to the detailed interpretation of this mechanism using Resonance Raman Spectroscopy. We first focus our attention at the interpretation of the spectra of the labels, i.e. model compounds [6], and then transfer this information to the actual enzyme.

2. Results and discussion

As proposed in the literature [5] the mechanism for photosensitivity and photodegradation of azoaldolase seems to consist in an isomerisation of some azocysteine residues from anti to syn configuration with a subsequent transfer of the azogroup from the cysteine to one of the histidines possibly lying topologically close. For the reason discussed above the study by Resonance Raman Spectroscopy of azoaldolase itself and of its model compounds related to the azo residues should provide specific information on the proposed mechanism.

In this paper we first discuss the information which can be derived mainly from the study of the model compounds which have been prepared just for this purpose. The series of models examined consists of various benzeneazo derivatives of N-acetylcysteine with different ring substituents in para position with respect to the azo group. The compounds studied can be described by the general formula:

where R=H, CH$_3$, OCH$_3$, COOH and NO$_2$. The more stable configuration of all these compounds is anti (E form). For R=NO$_2$ we have studied the vibrational spectrum of both anti and syn forms (E and Z forms respectively). For the case of histidine we have only obtained preliminary results on a mixture of isomers of p-carboxy-benzeneazo N-acetyl-L-histidine. The mixture of isomers is only related to the position of the azo substituent at the imidazole ring. The general formula is:

The striking selective character of the Resonance Raman Effect in these model compounds clearly results from examination of the vibrational spectra obtained from the Infrared and the Raman of N-acetylcysteine and benzeneazo N-acetylcysteine in the solid state (Cys and Azocys,respectively). We consider in more detail the case of these specific compounds as an example. Figure 1 shows the infrared absorption spectra of Cys and Azocys in the solid state. The large population of strong absorptions throughout the whole spectrum makes a detailed analysis very difficult for both compounds.

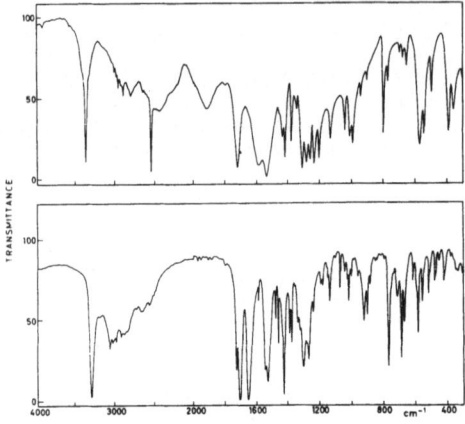

Fig.1 Infrared spectra from 4000 to 300 cm^{-1} of N-acetylcysteine (above) and Benzeneazo N-acetylcysteine (below) in solid phase (12 mm KBr pellet). Concentration: 1.8 and 1.2 mg, respectively; Slit width: 1.5 cm^{-1}; T.C.: 1.5 sec; Scan speed: 10 cm^{-1}/min.

An obvious qualitative interpretation of a few bands based on spectral correlations can be always done, but the isolation of a few bands structurally meaningful in the light of our work, is not straightforward. By comparing the Infrared spectrum of Azocys with that of Cys (Fig. 1) one observes that, in addition to the obvious disappearing of the S-H stretching at 2547 cm^{-1}, all bands which are characteristic of the amido-acid structure of Cys [7] are present, with small frequency shifts, in the infrared spectrum of Azocys. In the last substance these bands occur at 3304 (N-H stretching), 1705 (carboxyl C=O stretching), 1653 (Amide I), 1525 (Amide II) and 1298 (Amide III) cm^{-1}. Moreover, several additional bands appear in the spectrum of Azocys which must originate from vibrational modes localized in the azobenzene group. Among these, the two peaks of medium intensity at 766 and 685 cm^{-1} indicate monosubstitution at the benzene ring whereas the relative intensity of the doublet near 1600 cm^{-1} ($\sim$1600 shoulder, 1586 weak) suggests conjugation of the ring with an unsaturated group [8]. Several other peaks in the region from 1000 to 1600 cm^{-1} can be assigned to various ring stretching, ring deformation and C-H in-plane deformation modes of the monosubstituted aromatic ring [8]. Finally, the medium intensity feature at 1421 cm^{-1} falls in the region expected for the N=N stretching band. The asymmetric substitution around this group could well explain the relatively high infrared intensity of this vibrational mode.

From the above considerations one can conclude that the experimental infrared spectrum of Azocys is well in agreement with the chemical structure and that each vibrational mode gives rise to an infrared band. Due to the lack of any symmetry, the same result had to be expected in the Raman scattering experiments. If we turn to the Raman spectra (Fig. 2) one observes that practically for the case of Cys the bands in the infrared find their coincidences in the Raman spectrum. The lack of any symmetry element makes all normal

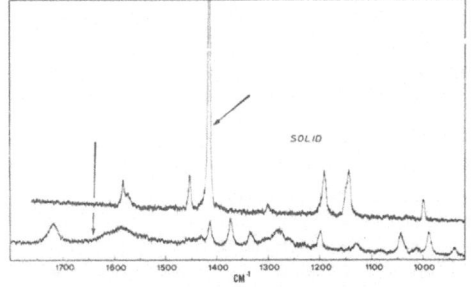

Fig.2 Raman spectra form 1750 to 950 cm^{-1} of N-acetylcysteine (below) and Benzeneazo N-acetylcysteine (above) in solid phase. Exciting line: Ar$^+$ laser, 4658 Å; Power on the sample: 1.1 mW; Slit width: 4 cm^{-1}; T.C.: 0.5 sec; Scan speed: 50 cm^{-1}/min.

modes infrared and Raman active even if the relative intensities are obviously different. On the contrary, the Raman spectrum of Azocys shows only a few bands thus indicating a strong simplification of the Raman spectrum if compared with the infrared. A partial assignment of the vibrational spectra of Azocys provides the way to distinguish which normal modes have disappeared in the Raman spectrum when compared with those observed in the infrared.

Examination of Figs. 1 and 2 shows that in the Raman spectrum of Azocys the vibrations originating from the cysteine residue are completely missing and that the nine Raman lines shown in Fig. 2 and two weak lines at 616 and 580 cm^{-1} originate from vibrational modes localized in the azobenzene group, i.e. in the chromophore [9]. Table I gives the vibrational assignment.

Table I   Vibrational assignment of the Raman spectrum of Benzeneazo N-acetyl-cysteine.

Raman		Infrared		
1594	m	1600	sh	} ring stretching
1586	vw	1586	vw	
1460	m	1459	m	ring stretching
1421	vs	1421	s	N=N stretching
1305	vw	1305	sh	ring stretching
1192	m	1190	w	aromatic CH deformation
1154	sh	1154	vw	aromatic CH deformation
1147	m	1147	w	C-N stretching
1000	w	1001	w	ring deformation
616	vw	616	sh	ring deformation
580	vw	580	m	ring deformation

The same kind of spectral pattern only arising from the chromophoric part of the molecule is found in the Raman spectra of all the other model compounds studied. It is particularly interesting to point out that the strongest peak at 1421 cm^{-1} in Fig. 2 is due to the stretching of the N=N group thus showing that such a group is strongly coupled with the electronic transition. Since the N=N is conjugated with the benzene ring the in-plane normal modes of the ring are intensified by their interaction with the electronic transition responsible for the intensity enhancement. The influence played on the properties of the N=N group by the electronic delocalisation of π electrons throughout the benzene ring is clearly shown when different groups are placed in para position. The frequency of the N=N stretching changes as well as its relative intensity with respect to the bands arising from the benzene ring.

$\nu_{N=N}$ for R=H, 1421; R=CH$_3$, 1433; = COOH, 1432; OCH$_3$, 1410 cm^{-1}. If the Raman spectra of solutions of these models in the same molar concentration in chromophoric groups as azoaldolase are recorded, the spectra are very weak with respect to the spectra of the solid, even if the same exciting line is used.

Next step is the analysis of the spectra of histidines. In Fig. 3 and 4 we show the Raman spectra of a mixture of azohistidines in the solid and H$_2$O solution. For these compounds the electronic spectrum shows an absorption near 450 nm. Since we excite the Raman effect with the lines of an Ar$^+$ Laser, the excitation under resonance conditions will be much more effective as clearly shown in Fig. 4. Even if a detailed vibrational assignment of these kinds of molecules is not yet available, we feel confident to propose that the resonance enhanced lines observed originate both from the benzeneazo group and from the imidazole ring.

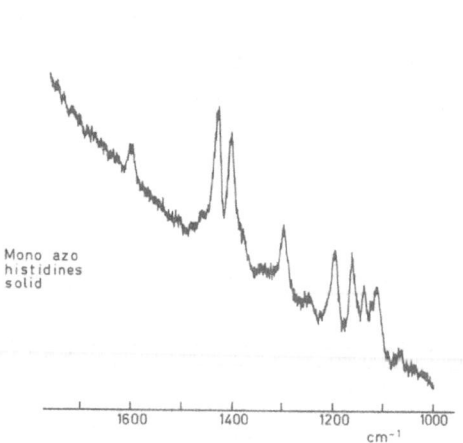

Fig.3 Raman spectrum form 1700 to 1000 cm^{-1} of p-carboxybenzeneazo N-acetylhistidines in solid phase. Exciting line: Ar$^+$ laser, 4658 Å; Power on the sample: 3 mW; Slit width: 4 cm^{-1}; T.C.: 0.5 sec, Scan speed: 50 cm^{-1}/min.

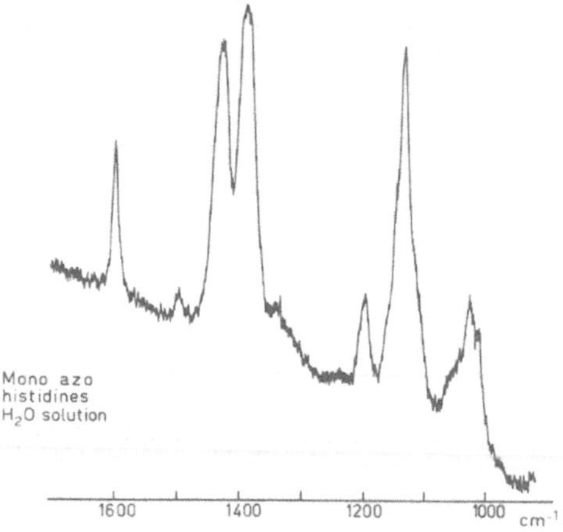

Fig.4 Raman spectrum from 1700 to 900 cm^{-1} of p-carboxybenzeneazo N-acetylhistidines in H$_2$O solution. Exciting line: Ar$^+$ laser, 4658 Å; Power on the sample: 40 mW; Slit width: 6 cm^{-1}; T.C.: 2 sec; Scan speed: 20 cm^{-1}/ min.

The three main lines around 1400 cm^{-1} and near 1150 cm^{-1} should become useful probes for the identification through Resonance Raman Spectroscopy of the existence of azohistidines residues in an otherwise complex protein network. More detailed studies of azohistidines are in progress.

After all these data have been collected from the study of model compounds we started the analysis of the spectrum of the labelled enzyme azoaldolase. The study is based on the comparison between the Raman spectra of the modified enzyme with that of the cysteine and histidine models in aqueous solution of comparable concentration of chromophores and with the same exciting line.

Figure 5 gives the electronic absorption spectrum of the modified enzyme and Fig. 6 reports the Raman spectrum of the same substance excited with the Ar$^+$ laser line at 4658 Å.

It is clear that under the experimental conditions chosen we are exciting far on the wing of the first absorption maximum at 330 nm. The general overall weakness of the Raman spectrum is due to the fact we are just approaching resonance conditions. In spite of the weakness of the spectrum reported in Fig. 6 it is still possible to recognise the features observed in the spectrum of the corresponding azocysteine model compound in aqueous solution in the same experimental conditions. There are no indications of the presence of a sizeable amount of azohistidine groups.

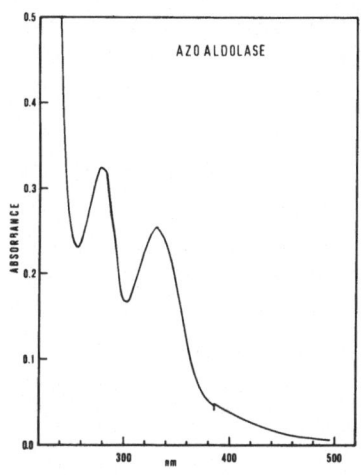

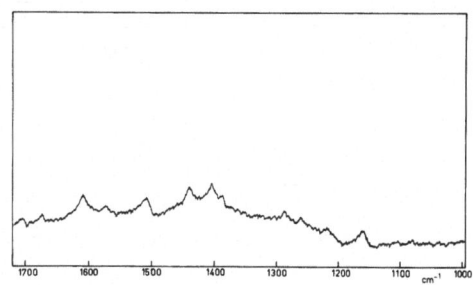

Fig.6 Raman spectrum form 1700 to 1000 cm^{-1} of H_2O solution of azoaldolase. Chromophore concentration: 1.6×10^{-4} M. Exciting line: Ar$^+$ laser, 4658 Å; Power on the sample: 20 mW; Slith width: 7 cm^{-1}; T.C.: 2 sec; Scan speed: 20 cm^{-1}/min.

Fig. 5 Absorption spectrum from 240 to 500 nm of azoaldolase in H_2O solution. Protein conc.: 0.283 mg/ml; Chromophore conc: 1.6×10^{-5} M.

This is the typical example of the type of selective conclusions which can be reached in these kinds of works. It becomes however evident the need of improving the experimental conditions of resonance excitation by approaching more closely the first absorption of the electronic spectrum of azoaldolase. However the excitation below 450 nm is not commonly reached by most of the conventional spectrometers and most of the sources of the laser light at present commonly used. More work is in progress in our laboratory.

From this study an observation of particular general interest in the use of labelled model compounds in resonance Raman studies of biological materials is the fact that the solid model compounds such as those studied in this work exhibit the characteristic behaviour of Resonance Raman excitation even when exciting lines away from the absorption maximum are used. We have tried several exciting lines and always observed this effect even with the red line of a Kr+ laser (6471 Å). Two possible reasons for this effect can be conceived. The first is that resonance excitation takes place even away from the absorption maximum because of the "natural" width of the absorption band in the solid. The second is the existence of additional electronic transitions of weak intensity which lie under the wings of the main band. The experimental fact that by slowly moving toward longer wavelengths with the exciting line the intensity enhancement does not significantly change seems to favour the first hypothesis.

187

3. Experimental

N-acetyl-L-cysteine from Serva (99% purity) was used without any further puri-fication. All the azo compounds were synthesized according to [4,6].

Visible and near U.V., infrared, and Raman spectra were recorded with a Perkin-Elmer mod. 356, Perkin-Elmer mod. 180 and Jarrel-Ash mod. 25-300 spectrophotometer, respectively. Coherent Radiation CR 3 Ar^+ and CR 500 K Kr^+ lasers were used to excite the Raman spectra; a 90° scattering geometry was chosen. The scattered radiation was revealed by a thermoelectrically cooled ITT FW 130 photomultiplier with photon counting device.

References

1. S. Sufrà, G. Dellepiane, G. Masetti and G. Zerbi, J. Raman Spectry. 6, 267 (1977)

2. P. R. Carey, in "Proceedings of the Fifth International Conference on Raman Spectroscopy" , Schulz Verlag Inc., Freiburg, 1976, and References herein

3. C. Y. Lai, N. Nakai and D. Chang, Science 183, 1204 (1974)

4. R. Felicioli, L. Nannicini, E. Balestrieri and G. Montagnoli, Eur. J. Biochem. 51, 467 (1975)

5. G. Montagnoli, S. Monti, L. Nannicini and R. Felicioli, Photochem. Photobiol. 23, 29 (1976)

6. G. Montagnoli, S. Monti and L. Nannicini, Gaz. Chim. Ital., 105, 559 (1975)

7. N. Fuson, M.L. Josien and R. L. Powell, J. Am. Chem. Soc. 74, 1 (1952)

8. L. J. Bellamy, "The infrared Spectra of Complex Molecules", 3rd edition, Methuen, London 1975

9. F.R. Dollish, W. G. Fateley and F. F. Bentley, "Characteristic Raman Frequencies of Organic Compounds", J. Wiley & Sons Inc., New York, 1974

Enhanced Anti-Stokes Raman Scattering from Living Cells of „Chlorella Pyrenoidosa"

F. Drissler

Max-Planck-Institut für Festkörperforschung, Heisenbergstraße 1,
D-7000 Stuttgart 80, Fed. Rep. of Germany

R.M. Macfarlane

IBM Research Laboratory, San Jose, CA 95193, USA

1. Introduction

Recent studies have shown that Resonance Raman Spectroscopy offers an interesting experimental method to determine the chemical and structural composition of systems even as complex as biological organisms. Its application for investigations of specific molecules in heterogeneous samples is therefore rather widespread [1-4].

In contrast to these approaches we have measured Raman spectra from living cells of the alga "Chlorella pyr." in an attempt to clarify a possible role of vibrational states for the biological process of photosynthesis. Since vibrations can be of importance for various kinds of mechanisms two different types of Raman investigations were used in order to study some aspects of the problem:

According to theoretical considerations by FRÖHLICH [5-7] a high non-thermal population can occur in living structures as a collective feature of the multi-component system if it is sufficiently supplied with energy. Since any population enhancement should lead to enhanced Anti-Stokes Raman scattering,one group of experiments was performed in order to measure the ratios R of corresponding Stokes and Anti-Stokes intensities. From the usual equations R is given as:

$$R(\Omega,\omega_L,T) = \frac{I_{as}^{\omega_L}}{I_s^{\omega_L}} = \frac{(\omega_L-\omega_E)^2(\omega_L-\omega_E-\Omega)^2}{(\omega_L-\omega_E)^2(\omega_L-\omega_E+\Omega)^2} \cdot P(\Omega,T) \qquad (1)$$

(Ω: vibrational frequency, ω_L: laser frequency, ω_E: absorption frequency of scattering particles, T: temperature)

The first factor in (1) expresses a cross-section term while P is the population factor which in the usual thermal equilibrium is just $n/(n+1)$. An analysis of P by the use of the measured values for R should clarify the question of a possible enhancement in population of vibrational states.

With a second kind of Raman studies we have compared spectra taken from cells with different photosynthetic activity in order to investigate changes which can be attributed to the different biological situation.

2. Materials and Methods

Suspensions of living cells of the green monocellular alga "Chlorella pyrenoidosa" were used for our experiments. Its photoactive pigments act in groups of about 300 molecules (photosystems) which are incorporated in special membranes. These units consist of Chlorophyll A (50 %), Chlorophyll B (16 %) and 7 slightly different kinds of Carotenoids (34 %). It has to be taken into account for evaluations of Raman intensities that these pigments have strong electronic absorptions in the visible range of the spectrum.

Our measurements were performed with a flow system to avoid heating of the cells with the laser. Due to a high flow speed [8] the algae were exposed to the laser light for times less than 2 msec. White background light was available for illumination of the reservoir in order to investigate relations between Raman intensities and the photosynthetic activity.

The results have been recorded using an Ar^+-laser for excitation as well as a 1m double monochromator and conventional photon counting detection for observation.

3. Experimental Results

A Stokes Raman spectrum of cells at room temperature is shown in Fig.1. The main features are 4 groups of lines (width ~15 cm^{-1})

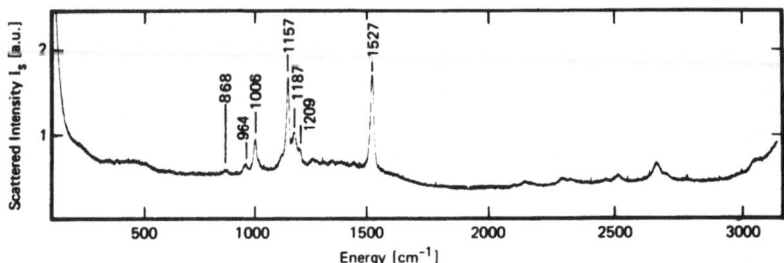

Fig.1 Stokes Raman spectrum observed at room temperature during irradiation with 514,5 nm (19436 cm^{-1}) without additional background light (laser power: 1,0 W).

between 800 and 1600 cm^{-1} which are superimposed on a rather structureless background. These lines are assigned to carotenoid vibrations by LUTZ [1].

The Anti-Stokes Raman spectrum for the same laser wavelength and power is shown in Fig.2 for a range from 800 to 1600 cm^{-1}.

As the laser wavelength was changed between 514,5 and 454,5 nm substantial changes in the scattered Stokes intensity occurred (Fig.3) which overwhelmed the non-resonant ω^4-behaviour and are attributed to resonances with the absorption band of carotenoid

190

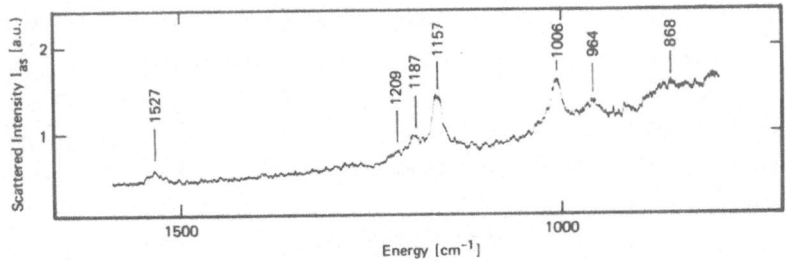

Fig.2 Anti-Stokes Raman spectrum observed under experimental conditions as described in Fig.1

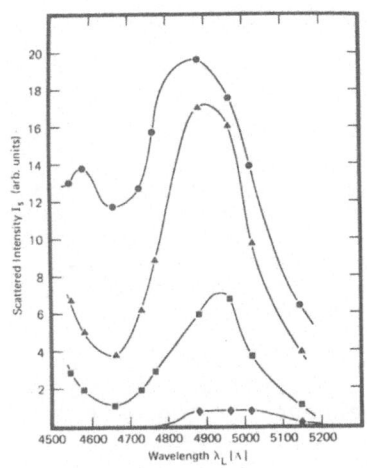

Fig.3 Intensity dependence of the Stokes shifted bands at 868 cm^{-1} (♦), 1006 cm^{-1} (■), 1157 cm^{-1} (▲) and 1527 cm^{-1} (●) on the wavelength of the incident laser light.

pigments at ~480 nm. As a striking result pronounced lines which can be attributed to chlorophyll molecules cannot be observed at all laser wavelengths.

The results as shown in Figs.1-3 were observed with cells kept in dark except for the time of exposure to the laser beam.

In an attempt to investigate the influence of changed photo-activity on the Raman spectra of the algae additional white light was used to illuminate the reservoir which is part of the flow system. This should keep the photoprocesses still active when the cells are not exposed to the laser light. Fig.4 shows the effect of additional white illumination on the Raman intensities at 1006 cm^{-1}:

- during additional illumination with white background light both, Stokes and Anti-Stokes intensity increase slowly

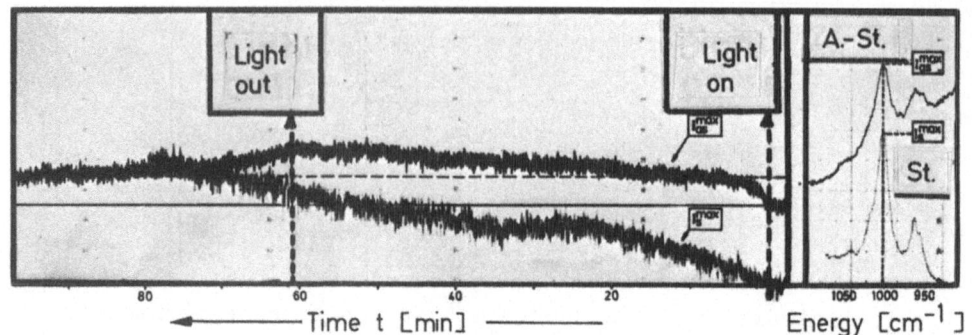

Fig.4 Influence of white background light on the Stokes (I_s^{max}) and Anti-Stokes (I_{as}^{max}) Raman intensities at 1006 cm^{-1}.

- the kinetic of the backgroundlight effect is different for Stokes and Anti-Stokes scattering which indicates a time-dependency of the ratio R

- the increase in intensity after one hour of illumination amounts to 13 % for the Stokes line and 31 % for the Anti-Stokes line

- the increase in Anti-Stokes intensity is partially reversible if background illumination is switched off. A reversibility of the increase in Stokes intensity was not observed.

It is important to note that an influence of additional white light was observed only if the cells were supplied with CO_2 at the same time. This shows clearly a dependence of the effect on the photosynthetic process.

4. Discussion

The intensity ratios R of the main Anti-Stokes bands (Fig.2) with respect to their Stokes counterparts (Fig.1) deviate considerably from the thermal population factor (Table 1). An investigation of the dependence of Anti-Stokes intensities on laser power showed that local heating by the laser is not responsible for the enhancement.

Furthermore, close to a resonance and in particular for large Ω the ratio R can be dominated by a frequency dependent cross--section factor which is different for Stokes and Anti-Stokes photons. We have attempted to separate this from the population factor P by considering the ratio of the Stokes intensities I_s^{ω} (laser at ω) and $I_s^{\omega+\Omega}$ (laser at $\omega+\Omega$) as given by Fig.3. For this case eq.(1) gives:

$$\frac{R}{\frac{n}{n+1}} = \frac{P}{\frac{n}{n+1}} \cdot \frac{I_s^{\omega+\Omega}}{I_s^{\omega}} \qquad (2)$$

Table 1 shows the enhancement factors which can be derived from our experiments with (2). It appears that the major part of the

Vibrational Energy $\Omega[cm^{-1}]$	Observed Ratio R $I_{as}^{19436}/I_{s}^{19436}$	Anti-Stokes Enhancement $R/n/(n+1)$	Res. Factor $I^{19436+\Omega}/I^{19436}$ From Fig. (3)	$\dfrac{P}{n/(n+1)}$	Table 1
868	0. 1	10.3	4.0	2.6±.5	
1006	0.060	7.2	5.6	1.3±.1	
1157	0.020	5.3	4.1	1.3±.1	
1527	0.0024	3.5	2.7	1.3±.2	

Anti-Stokes enhancement is due to a resonance with the carotenoid absorption band at ~ 480 nm. P appears to be enhanced by factors ranging from 1,3 to 2,6 if compared with the normal thermal equilibrium. Our data can, therefore, not rule out an enhancement in population. However, it is smaller than suggested by the theory mentioned at the beginning. In the present state our results do not indicate a high non-thermal population of vibrational states. On the other hand, we have to consider the biological conditions during our Raman experiment with respect to a strong photosynthetic activity of the cells since the predicted population enhancement depends on the function of the biological process.

An analysis of the different effect of additional white light on Stokes and Anti-Stokes intensities in terms of time yields a further increase of the ratio R.

5. Conclusions

Our investigations show that a great part of the measured Anti-Stokes enhancement is due to resonances in the scattered photon channel which are different for Stokes and Anti-Stokes photons. As a consequence, the analysis of resonance enhancements is of great significance for investigations of population effects.

Population factors derived from our experimental data indicate, however, that we cannot rule out some enhancement in population. The enhancement factors range from 1,3 to 2,6 if compared with the usual thermal population factor n/(n+1).

The observation of an influence of white background light on the Raman intensities demonstrates their dependence on processes related with the photosynthetic activity of the cell. Laser investigations, therefore, offer a sensitive method to study such processes.

References

1   M. Lutz, Biochim.Biophys.Acta, 460(1977)408
2   M. Lutz and J. Breton, Biochem.Biophys.Res.Comm., 53(1973)413
3   G.J. Thomas, The Spex Speaker, XXI,4(1976)
4   See e.g.: T.G. Spiro, Acc.Chem.Res., 7(1974)339
5   H. Fröhlich, Nature, 219(1968)743
6   H. Fröhlich, J.Quantum Chem., 2(1968)641
7   H. Fröhlich, Phys.Lett., 51A(1975)21
8   F. Drissler, W. Hägele, D. Schmid und H.C. Wolf, Z.Naturf.,
    32a(1977)88

Time Resolved Resonance Raman Scattering as a Probe of Hemoglobin Dynamics

J.M. Friedman and K.B. Lyons

Bell Telephone Laboratories
Murray Hill, NJ 07974, USA

1. Introduction

Hemoglobin (Hb) and myoglobin (Mb) have both been extensively studied; nevertheless, the structural and electronic basis for their reactivity and for differences in their reactivity is still uncertain. Various mechanisms have been proposed based upon static structural pictures derived from studies in which stable equilibrium configurations were probed. Implicit in any suggested mechanism is a sequence of structural and electronic events which can best be proved by monitoring these changes during the course of the reaction. For many Hb and Mb reactions of current interest, photodissociated carboxyhemeproteins provide an ideal system for studying structural and electronic events in real time.

Carbon monoxide and hemeproteins (Fe^{++}) form ligandprotein complexes that are photodissociated with high quantum efficiency by visible light [1]. By photolyzing the CO-protein complex with pulsed laser radiation it is possible to generate transient species whose properties reflect both nature of the photolytic and recombination processes as well as other nonequilibrium protein dynamics. Transient species generated by photodissociating carboxyhemoglobin (HbCO) have been heavily studied using transient absorption spectroscopy covering picosecond [2-4] nanosecond [5-8] and longer time scales [9-11]. These studies have been most effective in providing details about the time scale for photodissociation, diffusion limited recombination and quaternary structure rearrangements. However, the diffuseness of the absorption spectra makes it difficult to identify or characterize the various transient species involved in these processes. On the other hand, the Raman spectra of heme chromophores are spectrally sharp as well as having been characterized with respect to sensitivity to electronic and structural properties of the porphyrin. Using a single pulse [12] and double pulse [13, 14] techniques we have observed the resonance Raman spectrum of the transient heme species which occur 10 nsec to 1 msec subsequent to the photolysis of HbCO and MbCO. The significance and implications of these spectra are discussed in this paper.

2. Summary of Results

Using a single 10 nsec Nd: YAG laser pulse of 5320Å (~20mJ) to both photolyze and interrogate a solution of HbCO (10^{-4}M heme) we obtained [12] a resonance Raman spectrum of the transient species that occur within 10 nsec after photolysis. The resulting spectrum resembles that of deoxyhemoglobin; however, a careful comparison of the two spectra revealed that several of the Raman peaks associated with the transient were red shifted by as much as 5 ± 1 cm^{-1} relative to those of deoxy Hb. The transient spectrum contains no Raman peaks assignable to HbCO despite the favorable resonance conditions.

In the two pulse experiment, the tunable (frequency and time delay) output of a nitrogen laser pumped dye laser (10 n sec, $\leq$ 1mJ) was used to generate the Raman spectra of transients produced by an initial photolyzing pulse at 5320 Å (20mJ, 10 nsec). The dye laser was tuned into resonance with the Soret transitions of the hemes which results in a sizable enhancement of the polarized Raman peak at 1357 cm^{-1} for deoxy Hb, at 1377 cm^{-1} for HbO_2 and at 1373 cm^{-1} for HbCO. The frequency of this Raman transition is sensitive to the π electronic properties of the heme. As with the single pulse experiment, the deoxy Hb like Raman spectrum generated within a few nsec subsequent to photolysis exhibited shifted peaks (1353 cm^{-1} vis a vis 1357 cm^{-1}). Preliminary results obtained by comparing the spectra at several time delays indicate that at between .4 and 2 µsec subsequent to photolysis, this peak undergoes a blue shift towards the equilibrium value associated with deoxy Hb.

Using the double pulse system, we have compared [14] the recombination dynamics of photolyzed HbCO and MbCO. Firing the dye laser before the photolytic pulse, we find that in both systems, the spectrum reflects predominantly unphotolyzed material (photodissociation by the probe pulse has been minimized by focussing the beam to a line). Within the first few nsec after photolysis the HbCO and MbCO peaks have completely disappeared and a peak in the region of the deoxy species has appeared. This observation is consistent with previous results using a single pulse [12, 15, 16] and using CARS [17]. At later times a difference is observed in the two systems. In the Hb system the HbCO peak, at the expense of the deoxy Hb peak, regains nearly 50% of its intensity within 100 nsec of photolysis. On the same time scale no corresponding change is observed in the Mb spectrum. At much later times the expected recombination takes place in both Hb and Mb and the deoxy peak decays as the CO peak grows in. The temporal behavior of the Hb system is the same for excitation frequencies ranging from 4100 Å (favoring enhancement of HbCO spectra) to 4300 Å (favoring enhancement of deoxy Hb spectra) except that the peak ratios differ in the expected manner.

3. Implications with Respect to the Mechanism of Cooperativity

The hemes in HbCO are surrounded by a protein matrix that has the R quaternary structure whereas in deoxy Hb the protein has the T quaternary structure. Upon photolysis the switch from the R to T quaternary structure occurs on a time scale of tens of microseconds for the conditions employed in our experiments [11]. Consequently the shifts in frequency of the nsec transient spectra relative to the corresponding deoxy spectra may reflect the influence of the different protein structures upon the electronic properties of the heme. Alternatively these shifts may be an artifact of the photolytic pathway which could involve a long lived low lying electronic state. The observation that these shifts persist out to microseconds makes the latter explanation implausible. Recently, SHELNUTT et el [18] using Raman Difference Spectroscopy compared the Raman spectra of several deoxy Hb's, some of which were chemically modified so as to be stabilize the R quaternary structure. Those deoxy Hb's that were stabilized in the R

structure consistently yielded Raman spectra with several peaks shifted
($< 2cm^{-1}$) with respect to deoxy Hb's stabilized in the usual T structure.
The pattern of these steady state shifts is very similar to the shifts
observed in the nsec transient spectra which strongly implicates the effect
of the protein structure upon the heme as the origin of the shifts in the
time resolved spectra. However, shifts that appear in the time resolved
spectra appear to be at least a factor of two larger than those observed in
the steady state comparison. This difference in the magnitude of the shift
may be the result of variation of the tertiary structure of the protein
within a given quaternary structure. For equilibrated deoxy Hb's the tertiary structure of the protein is subject to the constraints of an unligated
heme, whereas for HbCO the heme is liganded. Therefore, the unligated heme
species generated within a few nsec subsequent to photolysis could still be
under the perturbing influence of an unstable tertiary structure of the
protein that has not yet relaxed into a configuration at equilibrium with
respect to the unligated heme. Consistent with the idea that tertiary
rearrangements may precede or trigger the R-T switch is the observation that
1353 cm^{-1} Raman peak of the nsec transient begins to shift on a time scale
that is at least an order of magnitude faster than the R-T switching time
and several orders of magnitude slower than any excited electronic state
relaxation time anticipated for a heme. These tertiary structure rearrangements which result from the ligation process, could well be the mechanism by
which the highly localized electronic and structural perturbation associated
with ligation propagates out to the structure of the protein matrix. We now
consider several such possible mechanisms in relation to the Raman data.

Movement of the iron upon ligation has been considered [19] as a probable
mechanical pathway by which the pertubation associated with ligation couples
to the protein. The shifted Raman peaks are all sensitive to electronic
properties of the heme which are responsive to the spin and structural state
of the iron, consequently, the shifting of these peaks on the .4 and 2 μsec
time scale could result from movement of the iron out of the porphyrin plane
in response to the loss of ligand. This explanation for the shifts is highly
implausible since motions of this kind should occur on a much faster time
scale especially when the protein is in the R structure which minimizes
protein associated motional constaints upon the iron. The fact that at 5 nsec
after photolysis the Raman peak known to be sensitive to both the spin state
of the iron and the porphyrin center to pyrrole nitrogen distance has a
frequency closely corresponding to unligated heme [12] is evidence but not
proof that on this time scale the iron has assumed an equilibrium position
with respect to the state of ligation.

A more plausible explanation based on the observed time scale for these
shifts is one in which the amino acids about the heme respond to the photo-
lytically induced alterations in the electronic and structural properties of
the heme. To account for the shifts observed in the Raman difference spectra
of the unligated Hb's, SHELNUTT et al [18] have proposed that the porphyrin
Π system can interact with the Π system of surrounding aromatic amino acids
such as Phe CD1. For deoxy Hb, the stronger this electronic interaction
the lower frequency of certain electronic marker bands such as the 1357 cm^{-1}
band. In the T structure this electronic interaction is minimized because
the geometry between the heme and the neighboring aromatic amino acids
is constrained by the rigid structure associated with the T quaternary
structure. However, for the deoxy Hb's having the R quaternary structure,

the corresponding geometry of the aromatic amino acids about the heme is determined by the stabilization forces associated with this electronic interaction which favors a greater degree of contact between the two π electron systems. Upon ligation (O_2, CO) the porphyrin ring experiences a loss of π electron density which, within this model, further increases the interaction with the aromatic amino acids. Consequently within a given quaternary structure, the energetics associated with the ligated heme favor a geometry in which the aromatic amino acid is closer to the porphyrin.

We now have the following scenario for the sequence of electronic and structural events which can account for the observed Raman shifts in the transient spectra. Our initial system is HbCO which is a fully liganded protein having the R quaternary structure. Because of these two factors, Phe CD1 (for example) is now closer to the porphyrin ring than in modified deoxy Hb stabilized in the R structure which in turn has a tighter geometry than ordinary deoxy Hb which has the T structure. Upon photolysis the electronic structure of the porphyrin, adjusts within a few psec, to the unligated state of the iron; however, the surrounding protein still has the now unstable structure of the fully liganded R species. Consequently the Raman spectra of the hemes immediately after photolysis manifest shifts that are larger than those observed for hemes contained in the modified deoxy Hb's where the protein has an equilibrated R structure with the hemes unligated. As Phe CD1 moves away from the porphyrin in response to the increased porphyrin π electron density, the appropriate Raman modes exhibit a concomitant shift towards frequencies associated with a deoxy Hb having a steady state R structure. This initial shifting should precede and may possibly herald the onset of the R to T transition. The final shifting of these Raman peaks to the position associated with normal deoxy Hb occurs when the equilibrium T structure is reached. Because of the pivotal position of the Phe CD1 linking the ligation site to the critical $\alpha_1\beta_2$ interface of the globin, this mechanism is very likely to be responsible for at least some fraction of the free energy of cooperativity in Hb.

4. Implications for the Origin of the Quantum Yield of Photolysis

Quantum yield (Q.Y.) measurements for the photolysis of HbCO and MbCO have previously been performed [20-22] using pulses of light of a µsec or longer duration. These studies indicate that MbCO has a nearly 100% Q. Y. whereas HbCO has a Q. Y. of about 50% for the conditions employed in our experiments. We find [14] that our initial 10 nsec pulse photolyzes 100% of the hemes in both the Hb and Mb samples; however, in Hb, 50% of the hemes undergo a fast geminate recombination with CO with a time constant of 65 nsec. Mb does not exhibit this fast recombination. Consequently, it would appear that this submicrosecond fast recombination is responsible for the variation in Q. Y. between Hb and Mb. Experiments are planned to determine whether chain specific variations in the heme cage is responsible for the observed effect in Hb.

References

1. J. Haldane and J. L. Smith, J. Physiol. *20*, 497 (1896).
2. C. V. Shank, E. P. Ippen, and R. Bersohn, Science *193*, 50 (1976).
3. L. J. Noe, W. G. Eisert and P. M. Rentzepis, Proc. Nat. Acad. Sci. USA *75*, 573 (1978).
4. B. I. Greene, R. M. Hochstrasser, R. B. Weisman and W. A. Eaton, Proc. Nat. Acad. Sci. USA *75*, 5255 (1978).

5. B. Albert, R. Banerjee and L. Lindqvist, Biochem. Biophys. Res Commun. *46*, 913 (1972).
6. B. Albert, R. Banerjee and L. Lindqvist, Proc. Nat. Acad. Sci. USA *71*, 558 (1974)
7. D. A. Duddell, R. J. Morris and J. T. Richards, J. C. S. Chem. Comm. *75* (1979).
8. B. Albert, S. El Mohsni, L. Lindqvist and F. Tfibel, Chem. Phys. Lett. *64*, 11 (1979).
9. R. H. Austin, K. W. Beeson, L. Eisenstein, H. Frauenfelder and I. C. Gunsalus, Biochemistry *14*, 5355 (1975).
10. N. Alberding, S. S. Chan, L. Eisenstein, H. Fraunfelder, D. Good, I. C. Gunsalus, T. M. Nordlund, M. F. Perutz, A. H. Reynolds and L. B. Sorenson, Biochemistry *17*, 43 (1978).
11. C. A. Sawicki and Q. H. Gibson, J. Biol. Chem. *251*, 1533 (1975).
12. K. B. Lyons, J. M. Friedman and P. A. Fleury, Nature *275*, 565 (1978).
13. J. M. Friedman and K. B. Lyons, Proceedings of the Second US-USSr Light Scattering Symposium ed. H. Cummins (Plenum Press 1979) in press.
14. J. M. Friedman and K. B. Lyons submitted for publication.
15. R. B. Srivastawa, M. W. Schuyler, L. R. Dosser, F. J. Purcell and G. H. Atkinson, Chem. Phys. Lett. *56*, 595 (1978).
16. W. H. Woodruff and S. Farquharson, Science *201*, 831 (1978).
17. R. F. Dallinger, J. R. Nestor and T. G. Spiro, J. Amer. Chem. Soc. *100*, 6251 (1978).
18. J. A. Shelnutt, D. L. Rousseau, J. M. Friedman and S. R. Simon, Proc. Nat. Acad. Sci. USA *76*, 4409 (1979).
19. M. F. Perutz, Nature *228*, 726 (1970).
20. R. W. Noble, M. Brunori, J. Wyman and E. Antonini, Biochemistry *6*, 1216 (1967).
21. W. A. Saffran and Q. H. Gibson, J. Biol. Chem. *252*, 7955 (1977).
22. C. A. Sawicki and Q. H. Gibson, J. Biol. Chem. *254*, 4058 (1979).

Direct and Reverse Photoreactions in Bacteriorhodopsin at Low Temperatures on Picosecond Time Scala

P.G. Kryukov, Yu.A. Matveetz, and A.V. Sharkov

Institute of Spectroscopy, Academy of Sciences of the USSR, Moscow Region Troitzk 142092, USSR

Yu.A. Lazarev and E.L. Terpugov

Institute of Biophysics, Academy of Sciences of the USSR, Moscow Region Poushchino 142292, USSR

1. Introduction

Bacteriorhodopsin (BR) is the only protein in the purple membrane of Halobacterium halobium [1]. Like visual pigment rhodopsin, it contains retinal chromophore bonded to protein via a protonated Schiff base linkage [1,3]. This results in an intense absorption band of purple membrane with its maximum at 570 nm [1]. The light action causes a number of subsequent chemical transformations in BR culminated in the proton transfer across membrane and return of BR to its initial state, designated as BR570 [2]. BR acts as a light-driven proton pump and the light energy is stored up as the energy of transmembrane potential difference [4].

The primary step of the photochemical cycle in BR is the formation of the photoproduct K610 with the maximum of absorption band at 610 nm [2,5]. At room temperature within several microsecond K610 transforms to the next intermediate of the photochemical cycle. This reaction needs no light. At a temperature below -120°C K610 is stable. Under these conditions the continuous illumination of BR570 is followed by a red shift of the absorption band. The related difference spectrum has the maximum at 635 nm [2,5].

Under picosecond excitation there was an increase in absorption at 635 nm within the 6 psec pulse duration [6]. The measurements at -150°C did not reveal other kinetics during at least several seconds [7]. It was shown by the use of subpicosecond technique that the changes in absorption at 615 nm arise within 1±0.5 psec [8]. At 580 nm rapid bleaching (<6 psec) is followed by increasing absorption within about 15 psec at room temperature [6] and 20 psec at 68°K [9]. The measurements of difference spectrum with picosecond resolution over a wide spectral range reveal transient state preceding K610 and formed during the excitation time [10,11]. At room temperature the K610 formation time is 11 psec [10]. It rises with temperature decrease and deuteration of labile hydrogens in bacteriorhodopsin [10]. In the region 610-640 nm the transition from the observed intermediate state to K610 is characterized by small changes in absorption within the limits of the experimental error. The transient arises probably during 1±0.5 psec.

The photochemical reaction BR570→K610 is reversible, i.e. the process K610→BR570 also takes place as a result of light absorption by K610 [5]. The information on the relation between the rates of the direct and reverse reactions is essential when the physical mechanisms of the primary photochemical events in bacteriorhodopsin are considered. The present paper deals with the kinetic study under picosecond excitation of the two above mentioned photoreactions in light-adapted [2] deuterated bacteriorhodopsin at 13°K:

$$BR570 \rightleftharpoons K610$$

The rates of the direct and reverse reactions are found unequal. The back reaction is much faster than the direct one. It is suggested that the BR570→K610 and K610→BR570 phototransitions are characterized by different reaction ways.

2. Materials and Methods

Purple membranes extracted and deuterated by conventional methods [12] were suspended in the ethylene glycol - D_2O (2:1 v/v) mixture. The concentration of purple membrane was 0.4 mMol/L. The absorbance in the 2 mm cell at 570 nm was about $A_{max}=2$ O.D. The low temperature spectra were obtained with a standard helium cryostat. The kinetic measurements were carried out at 13°K with a double beam picosecond spectrometer (Fig.1).

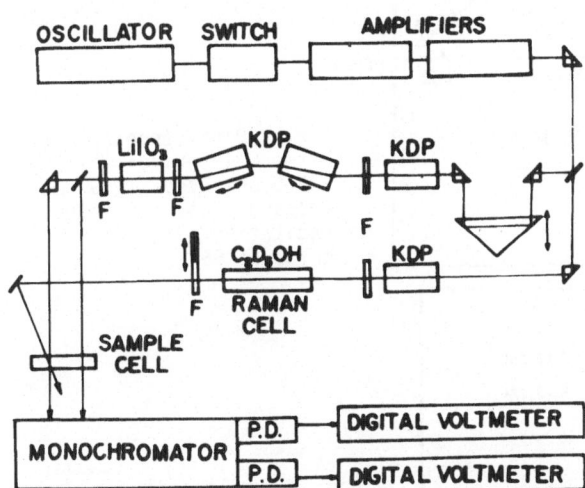

Fig. 1 Experimental set-up.

The direct photoreaction BR570→K610 was excited by Nd-YAG laser pulses at 532 nm and 1 mJ energy. The absorption changes dependence on the exciting pulse energy was linear [7]. Along with 532 nm laser pulses for the excitation of the direct photoreaction we used 630 nm light pulses. 630 nm is the frequency of the second Stokes component of the stimulated Raman scattering of 532 nm pulses in a 10 cm ethanol cell. The reverse photoreaction K610→BR570 was excited by 689 nm light pulses

of the second Stokes component in deuterated ethanol (C_2D_5OH).
BR570 has no absorption at 689 nm. Changing filters behind
Raman cell we excited direct or back reactions. The second-
-harmonic pulses of a single-path parametric oscillator on KDP
crystals were used as probe radiation [13]. The parametric
oscillator was pumped by 532 nm pulses. Such method provided
pulses of a more stable energy and lesser divergence compared
to the picosecond continuum. The probe radiation was detected
with germanium photodiodes and digital voltmeters.

To define the start-time point and pulse duration, the cell
in the picosecond spectrometer was replaced by KDP crystal
where the sum-frequency signal was induced as a result of nonlinear interaction between the 1064 nm pulse of the known duration (30 psec) and the pulse at 532 nm or 689 nm (the optical path of 1064 nm pulse coincided with the optical path of
probe pulse). The start-time point and pulse durations at
532 nm and 689 nm were determined from the dependence of the
sum-frequency signal on the optical delay between the probe
and excitation pulses (Fig.4A). The essentially different

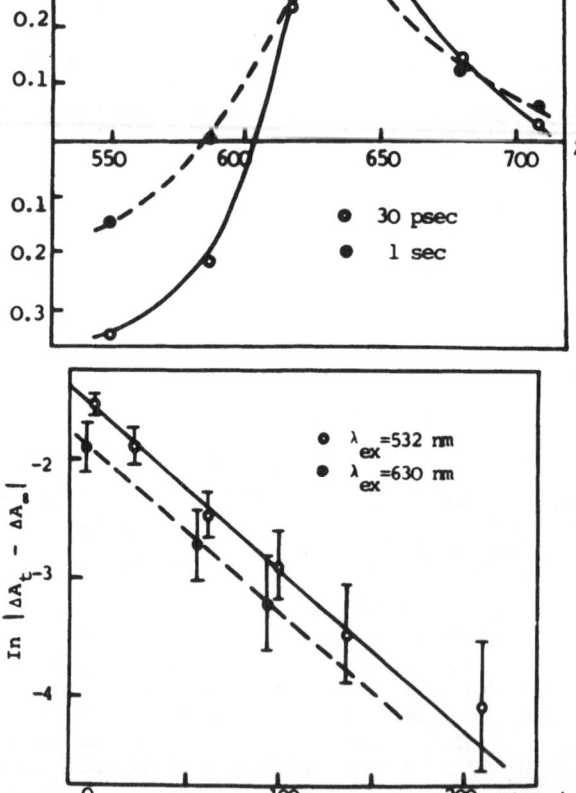

Fig. 2 Difference spectra
between intermediates and
BR570 observed at 30 psec
(o) and 1 sec (●) after
excitation with 30 psec
pulse at 532 nm.

Fig. 3 Semilogarithmic plots
of the BR570 K610 transition
observed at 590 nm under 532
nm (o) and 630 nm (●) excitation. ΔA_∞ - absorption
changes at 1 sec after excitation. Zero point corresponds to the moment of
completion of exciting pulse
action.

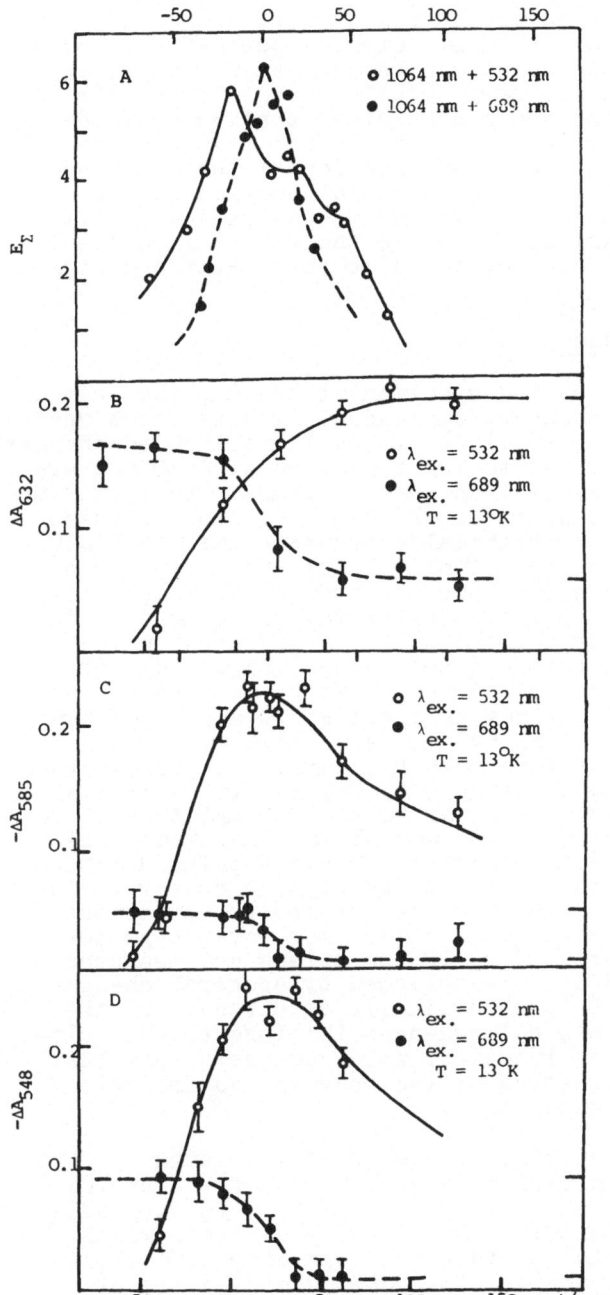

Fig. 4 A-The dependence of the sum frequency signal on the optical delay between probe (1064 nm) and exciting pulses at 532 nm (o) and 689 nm (●). B,C,D-kinetics obtained at 632 nm (B), 585 nm (C) and 548 nm (D) following excitation of BR 570 at 532 nm (o) and excitation of K610 at 689 nm (●). Lower time scale is for the direct reation, upper one for the back reaction.

203

forms of two correlation curves in Fig.4A were connected with
the depletion of the central (most intense) part of the second-
harmonic pulse due to the Raman scattering. As a result, the
pulse at 532 nm with the primary duration about 30 psec (without Raman cell) was broadened and acquired a structure. The
correlation function for the 689 nm pulse had a 40 psec half-
-width, corresponding to the pulse duration of the Raman scattering signal no more than 30 psec. The distortion of the
second-harmonic pulse led to the 20 psec divergence of the
correlation function maxima. Therefore the starting points in
the time scale in Fig.4 did not coincide for direct and back
processes.

3. Results and Discussion

During the pulse action on the specimen at 532 nm, the absorption spectrum of BR570 changes. Corresponding difference spectrum is shown in Fig.2 (solid line). These light-induced changes
differ from the ones related to the formation of K610 and are
registered 1 sec after excitation (dashed line). The formation
of stationary difference spectrum (K610-BR570) from transient
one is described by the exponential dependence with τ =80±10 ps.
The kinetics is shown in Fig.3 (solid line). The sample was
excited by 30 psec pulses (not passed through the Raman cell)
and the kinetics was investigated after the pulse action. As
the sample were excited by a light pulse of different wavelength (630 nm) the kinetics did not change (Fig.3, dashed
line). The formation time of K610 seems to be independent on
pump frequency. The solid lines in Fig.4 B,C,D show the changes
in absorption at 632, 585 and 548 nm when BR570 was irradiated
by a pulse at 532 nm passed through the Raman cell. The pulse
duration in this case is comparable to the formation time of
K610. However, the kinetics similar to that recorded at 30 psec
excitation can be observed at 548 and 585 nm. The rise time
measured at 632 nm correlates with the excited pulse duration.
There were no further changes in the absorption observed. The
process of K610 formation was not accompanied by variations
in absorption at this wavelength [6-11]. The reverse photoreaction K610 → BR570 excited by 689 nm pulses and measured
at 632 nm, 585 nm and 548 nm is followed by spectral changes
presented in Fig.4 B,C,D (dashed lines). As one can see the
transition time at all probe frequences is close to the half-
-width of the correlation function, which suggests very fast
spectral changes. The kinetics of the reverse photoreaction
is not resolved in our experiment.

The observed transient state preceding the arising of K610
may be interpreted as an excited electronic state of BR570.
We suppose that the process giving rise to K610 is the radiationless electronic relaxation into the ground state which
has a minimum of potential surface corresponding to K610. As
a result of relaxation process cis-trans isomerisation of
retinal or other structural conformations are possible [14].
The quantum yield of photoproduct formation is φ=0.3 [15].
The rest particles relaxe without radiating back to BR570
(the quantum efficiency of fluorescence is negligible [16]).
The way of photoproduct formation does not differ, in principle,

from that of particle return to the initial state. The observed
kinetics of the K610 formation (decay of the transient state)
corresponds to electronic relaxation from the minimum of the
excited electronic state potential surface into the ground
state. The vibrational relaxation time in the excited state
(or relaxation time from more excited electronic states) is
negligibly small as follows from the independence of the transition kinetics on the frequency of action light. One may assume
that this time equals to ⁓1 psec at 615 nm light excitation.
Rate of the K610 formation reveals the Arrenius temperature
dependence at high temperatures and is practically temperature-
-independent at low temperatures [10]. For electronic relaxation
processes the dependence of such a type can take place when
potential surfaces of the excited and ground state are inter-
sected near to the minimum of the upper state [17].

The sum of the quantum yields of direct and back reactions
is equal to 1 [15]. It means that there is a transient state
common in both photoprocesses [18]. It is possible to suggest
that the excited electronic state has a single minimum (along
reaction coordinate) which is reached by exciting either BR570
or K610 [18]. If the reaction rate is determined by the decay
of this common intermediate state, then it does not depend on
the product to be excited: BR570 or K610. In this case the
process of returning into the initial state (BR570) under K610
excitation should be described by 80 psec kinetics as a direct
process. As we didn't observe the analogous kinetics then we
dare say that the reverse reaction in faster than the direct one.

As the times of the direct and reverse photoreactions do not
concide it is reasonable to suppose that the potential surface
of the excited electronic state has two minima separated by a
potential barrier preventing transition between them (Fig.5).
The electronic relaxation from the 1-st minimum (when BR570 is
excited by light) and from the 2-nd minimum (for the K610 →
BR570 transition) is assumed to populate excited vibrational
levels of the ground electronic state. These levels can be

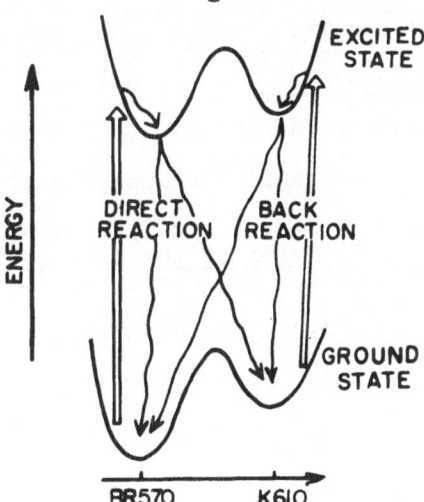

Fig. 5 Scheme of the primary events in
bacteriorhodopsin.

common for both processes if they are situated higher than the
potential barrier between K610 and BR570 in the ground state.
Further vibrational relaxation results in the formation of the
BR570 and K610 with appropriate yields.

Acknowledgments

We wish to express our deep appreciation to Prof. V.S.Letokhov
who showed great interest in the work and also to L.N.Chekulaeva
who supplied us with specimens.

References

1. D.Oesterhelt and W.Stoeckenius, Nat.New.Biol. _233_, 149
   (1971).
2. W.Stoeckenius and R.H.Lozier, J.Supramol. Struc. _2_, 769
   (1974).
3. A.Lewis, J.P.Spoonhower, R.A.Bogomalni, R.H.Lozier, and
   W.Stoeckenius, Proc.Natl.Acad.Sci. U.S.A. _71_, 4462 (1974).
4. E.Racker and W.Stoeckenius, J.Biol.Chem., _249_, 662 (1974).
5. R.H.Lozier, R.A.Bogomolni, and W.Stoeckenius, Biophys.
   J._15_, 955 (1975).
6. K.J.Kaufmann, P.M.Rentzepis, W.Stoeckenius, and A.Lewis,
   Biochem.Biophys.Res.Comm. _68_, 1109 (1976).
7. P.G.Kryukov, Yu.A.Lazarev, V.S.Letokhov, Yu.A.Matveetz,
   E.L.Terpugov, L.N.Chekulaeva, and A.V.Sharkov, Biofizika
   (Biophysics) _23_, 171 (1978).
8. E.P.Ippen, C.V.Shank, A.Lewis, and M.A.Marcus, Science
   _200_, 1279 (1978).
9. K.J.Kaufmann, V.Sundstrom, T.Yamane, and P.M.Reutzepis,
   Biophys.J. _22_, 121 (1978).
10. M.L.Applebury, K.S.Peters, and P.M.Reutzepis, Biophys J.
    _23_, 375 (1978).
11. P.G.Kryukov, Yu.A.Lazarev, Yu.A.Matveetz, E.L.Terpugov,
    and A.V.Sharkov, Izv.Acad.Nauk SSSR _43_, 1498 (1979).
12. A.Oseroff and R.Callender, Biochem. _13_, 4243 (1974).
13. S.A.Akhmanov, A.Yu.Borisov, R.V.Danelius, A.S.Piskarskas,
    A.P.Razjivin, and V.D.Samuilov, Pisma v JETF _26_, 655 (1977).
14. G.S.Hammond- In: Advance in photochemistry. v.7, N.Y.-L.,
    1969, p.373.
15. C.R.Goldschmidt, O.Kalisky, T.Rosenfeld, and M.Ottolenghi,
    Biophys.J. _17_, 179 (1977).
16. A.Lewis, J.P.Spoonhower, and C.J.Perreault, Nature (Lond.)
    _260_, 675 (1976).
17. V.L.Ermolaev, E.N.Bodunov, E.B.Sveshnikova, and T.A.Shakh-
    verdov - In: Radiationless energy transfer of the electro-
    nic excitation, ed. M.D.Galanin, Nauka - Leningrad, 1977,
    p.204 (Russian).
18. T.Rosenfeld, B.Honig, M.Ottolenghi, J.Hurley, and T.G.
    Ebrey, Pure and Appl.Chem. _49_, 341 (1977).

Selective Action on Nucleic Acids Components by Picosecond Light Pulses

D.A. Angelov, P.G. Kryukov, V.S. Letokhov, D.N. Nikogosyan, and A.A. Oraevsky

Institute of Spectroscopy, USSR Academy of Sciences, Moscow region
Troitzk 142092, USSR

The development of modern laser techniques provided wide range
of variation of radiation parameters such as frequency, intensity, pulse duration, thus making it possible to carry out
investigations on selective action on the substance. The term
" selective action " in our understanding means that molecules
or parts of molecules of the same type undergo considerable
change (modification) under laser irradiation [1]. Such change
can be caused by photoionization or photodissociation with
subsequent chemical reactions.

This investigation devoted to the problem of selective action
on nucleic acids components is aimed at achieving the modification of the bases of the same type in native DNA and RNA. This
is important for the development of physical approaches to
nucleic acids sequencing [2]. Successful solution of such a
problem would provide investigators with a fine delicate method
of action on complex biological molecules.

This work reports about the first results proving the fact
that it is possible to carry out selective action on one of the
nucleic acids bases in comparison with the other bases. Selective action is achieved both in the case of irradiation of
mixtures of bases and in the case of irradiation of mixtures of
more complex components such as nucleosides and nucleotides.
We have also irradiated dinucleotide which is a part of single
stranded RNA and carried out selective action on one of two
bases included into it.

The realization of selective action on nucleic acids bases
faces principal difficulties. First of all it is necessary to
note that linear UV absorption spectra of nucleic acids bases
show little difference. There are other serious difficulties
such as fast relaxation of excited electronic states and fast
intramolecular excitation transfer [3]. These difficulties can
be overcome in the following ways. Because of little difference
in linear UV absorption spectra we can use two-step (or multi-
step) excitation of high lying electronic states through the
intermediate singlet state S_1. In this case both photoionization
and photodissociation with subsequent chemical reactions may
take place. The photoproducts yield in such a stepwise photoprocess depends on a large number of parameters such as electronic levels structure, intensity of irradiation, lifetime of
intermediate electronic state, absorption cross-sections from
the ground and excited intermediate states, quantum yield of
chemical reactions from high lying electronic states. So we can

hope to achieve great differences in photoproducts yield for different bases, i.e. to carry out selective action on any chosen base.

Experimental implementation of these considerations requires a powerful picosecond tunable UV radiation source for effective population of high lying electronic states. The properties of excited electronic states as well as the quantum yields of subsequent chemical reactions depend on the solution pH. The loss of selectivity due to intramolecular excitation transfer can be considerably decreased by using rather short laser pulses ww for excitation.

In this work the picosecond set-up [4-6] was used for excitation of the high lying electronic states of nucleic acids components. Due to harmonic generation, stimulated Raman scattering and sum-frequency generation it was possible to obtain single picosecond pulses (30ps duration) with the energy 15mJ on the wavelength 266nm, with the energy 3mJ on the wavelengths 269nm, 273nm, 281nm, 289nm, 300nm, 315nm and with the energy 0.1mJ on the wavelengths 218nm, 222nm, 227nm, 234nm, 244nm. In most experiments discussed below we used pulses with the wavelength 266nm which is near the maximum of the first electronic UV absorption band of nucleic acids components. The pulse repetition rate was 0.5Hz. The laser beam cross-section area was 0.28cm^2. The volume of the solution under irradiation was 0.75cm^3, the cell volume was 1cm^3, the cell thickness was 0.5cm. The concentration of the solutions irradiated was $10^{-4} \div 10^{-5}$M. The solution in the

a)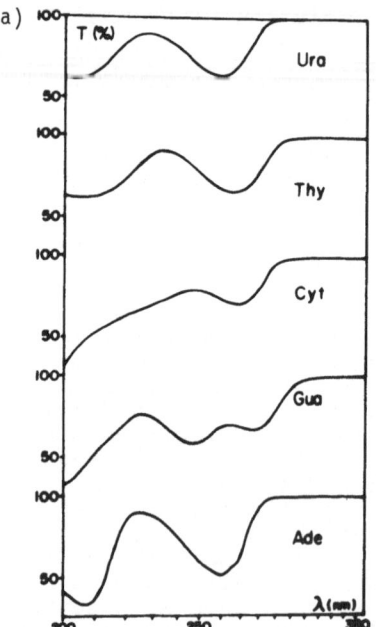

b)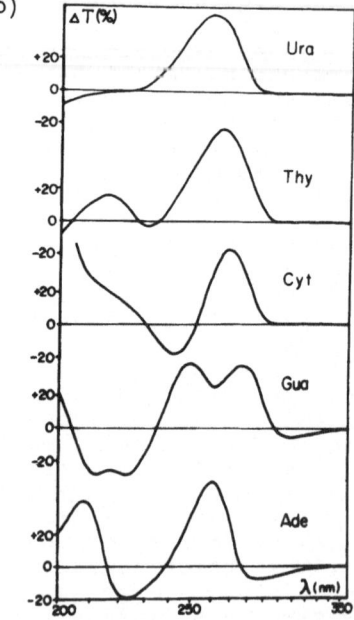

Fig.1 UV absorption spectra for all five nucleic acids bases in neutral water solution (pH = 6.3) (a) linear UV absorption spectra (b) difference UV absorption spectra taken after irradiation

quartz cell was continuously mixed during irradiation without access of air.

We have recently shown that when diluted aqueous solutions of nucleic acids bases are irradiated by UV pulses with the wavelength 266nm and intensity 1Gw/cm^2 two-step photochemical reactions occur [7]. As a result of these reactions irreversible photoproducts are formed. It should be noted that the difference UV absorption spectra taken after irradiation in contrast with linear ones have a characteristic shape for each of nucleic acids bases [8] (Fig.1). When the mixture of bases is irradiated the difference spectrum of photoproducts obtained is the sum of difference spectra of separate bases. This circumstance provides a simple and convenient method for evaluating the selectivity of action on various bases.

First we tried to carry out selective action on one of the two bases in equimolar water mixture. In order to find the pair of bases for which the difference in photoproducts yields is maximum we have irradiated consequently all the five bases of nucleic acids in neutral water solution (pH = 6.3). Fig.2a shows the dependences of the value $\Delta D/DE$ versus the intensity of irradiation for all five bases. This value (we shall call it " photoproducts yield ") is determined by the optical density relative change in the maximum of the difference spectrum per irradiation dose. The irradiation dose was calculated taking into account the fact that simultaneously only a part of molecules in the cell was irradiated. The linear dependence of the curves $\Delta D/DE$ at small intensities confirms the two-step character of the observed optical density change. The bend of the curves $\Delta D/DE$ corresponds to the saturation of electronic transition $S_0 \rightarrow S_1$ and makes it possible to obtain the upper estimate of the lifetime τ_1 of the electronic state S_1. If we consider the UV pulse duration to be equal to 30ps, then according to [7] we obtain $\tau_{1Cyt} = 36\beta ps$, $\tau_{1Ura} = 35\beta ps$, $\tau_{1Gua} = 31\beta ps$, $\tau_{1Thy} = 26\beta ps$, $\tau_{1Ade} = 9\beta ps$, where $\beta = 1/(1 + \sigma_1/\sigma_2)$, σ_1 and σ_2 are absorption cross-sections from the ground (S_0) and the excited (S_1) electronic states for each of the bases. The plateau of the curves $\Delta D/DE$ allows us

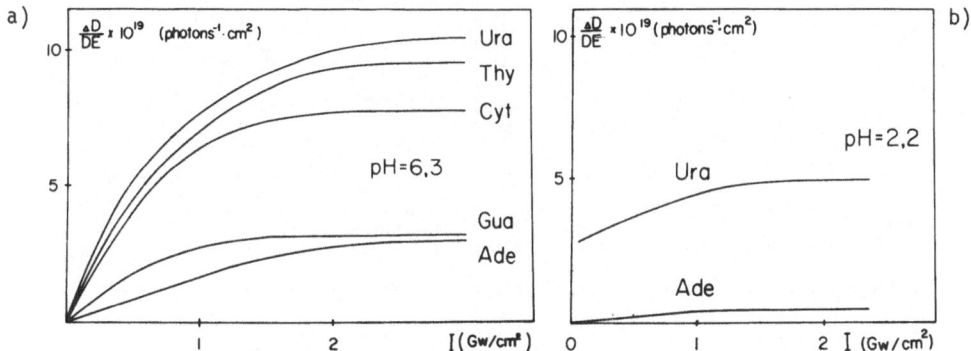

Fig.2 The dependences of photoproducts yields versus irradiation intensity (a) for all five bases in neutral aqueous solution (b) for uracil and adenine in acid aqueous solution. The irradiation wavelength $\lambda = 266$nm

to obtain the lower estimate of the quantum yield φ of two-step photoreactions. Again according to [7] we have $\varphi_{Ura} = 8\gamma\%$, $\varphi_{Cyt} = 6.7\gamma\%$, $\varphi_{Thy} = 6.3\gamma\%$, $\varphi_{Gua} = 2.3\gamma\%$, $\varphi_{Ade} = 1.3\gamma\%$, where $\gamma = 1/\beta$.

For experiments on selective action we have chosen the pair of bases uracil - adenine, which shows maximum difference in photoproducts yields in neutral water solution. For further increase of this difference the optimization of the solution pH was carried out. As it turned out photoproducts yields of uracil and adenine in acid aqueous solution (pH = 2.2) differ more (Fig.2b). In this case we obtain the following estimates: $\tau_{1Ura} = 23\beta ps$, $\tau_{1Ade} = 14\beta ps$, $\varphi_{Ura} = 3.8\gamma\%$, $\varphi_{Ade} = 0.2\gamma\%$.

Note that the curve for uracil does not approach to zero when the intensity is decreased. This points to the fact that a part of products of UV irradiation of uracil (in our case $\sim 30\%$) is the result of one-step photoreactions. It is natural to assume that these photoproducts correspond to the well-known uracil hydrate [9]. The estimates of the quantum yield ($\varphi = 1.4\%$) and of the reversion time of one-step photoproduct ($\tau = 180$min at $t^o = 27^oC$) correlate with the data for uracil hydrate [9] that confirms the assumption proposed.

Fig.2b also shows that it is possible to choose such irradiation intensity at which the difference in photoproducts yields is maximum. The selectivity of action on uracil relative to adenine is determined by the ratio of photoproducts yields and equals

$$S = (\Delta D_{Ura}/D_{Ura}) \, / \, (\Delta D_{Ade}/D_{Ade})$$

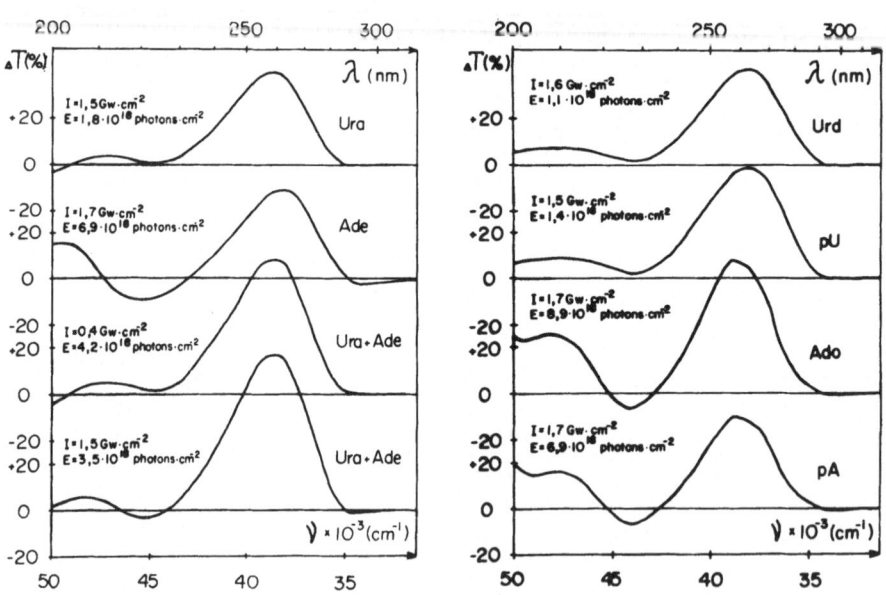

Fig.3 Difference spectra of photoproducts obtained after UV irradiation of acid aqueous solutions (pH = 2.2) (a) of uracil, adenine and their equimolar mixture (b) of uridine, uridine - 5´- phosphate, adenosine, adenosine - 5´- phosphate

We have irradiated the equimolar aqueous mixture uracil -
adenine ($C_{Ura} = C_{Ade} = 4 \cdot 10^{-5}M$, pH = 2.2) at two intensities.
The difference spectra of photoproducts obtained are shown in
Fig.3a. At the irradiation intensity 0.4Gw/cm^2 the difference
spectrum of mixture practically coincides with the difference
spectrum of uracil photoproducts. The calculated selectivity
of action on uracil determined from Fig.2b in this case is
equal to 22.3. At the irradiation intensity 1.5Gw/cm^2 in the
difference spectrum of mixture bands of adenine photoproducts
with λ = 218nm and λ = 295nm appear. Simultaneously a small
shift of the difference spectrum maximum to the long-wave region is observed. The selectivity of action on uracil determined
from the difference spectrum of mixture photoproducts is 11,
that agrees well with the calculated value 11,3 obtained from
Fig.2b. This points to the fact that there are no additional
losses of selectivity in the mixture under our conditions. It
is worth noting that the UV absorption spectrum of the solution
after continuous irradiation within the region 255 ÷ 285nm is
identical to the absorption spectrum of adenine.

Thus in spite of very small difference in the linear UV
absorption spectrum we have managed to obtain an essential
difference in photoproducts yields for uracil and adenine in
acid aqueous solution and in this way to carry out laser selective action. In this case the selectively modified base was the
pyrimidine base (uracil) which is photochemically less stable
than the purine base (adenine). However it should be noted
that the methods we have in our disposal can provide selective
action on a purine base which is in mixture with a pyrimidine
one.

To prove the possibility of the selective modification of the
purine base we have irradiated all five nucleic acids bases at
the wavelength 289nm. On this wavelength the absorption of one
purine base - guanine - considerably exceeds the absorption of
other nucleic acids bases, thus leads to great difference in
photoproducts yields. Fig.4 shows experimental dependences of
photoproducts yield versus irradiation intensity for all five
bases in neutral aqueous solution. The selectivity of action on
guanine relative to uracil is S = 13, to cytosine is S = 17, to
thymine is S = 50, to adenine is S = 80. It should be emphasized
that in this case the saturation of $S_0 \rightarrow S_1$ transition is not
achieved. A further increase of intensity gives rise to two-photon absorption and thus to the loss of selectivity.

In this experiment we have proved the possibility of selective action on guanine due to two-step photochemical reactions
under powerful single-frequency irradiation. In order to carry
out selective action on any desired base we must use the difference in absorbtion cross-sections from the excited electronic state S_1 and thus apply two-step excitation by two pulses
with different frequencies. By introducing a proper delay
between pulses it is also possible to utilize the difference
in lifetimes τ_1 of various bases and thus to increase more the
degree of selectivity.

Our further efforts were directed to the realization of
selective action on the nucleic acids bases which are contained
in nucleosides and nucleotides. Since the UV absorption properties of nucleosides and nucleotides in the region of wave-

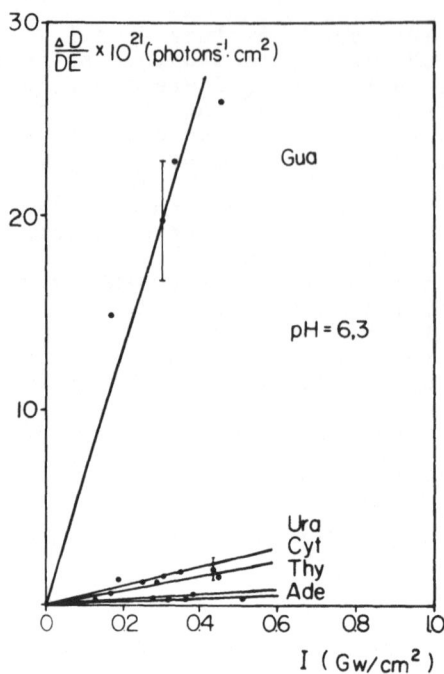

Fig.4 The dependences of photo-products yield on irradiation intensity for all five bases in neutral aqueous solution. The irradiation wavelength λ = 289nm

lengths more than 230nm are determined by UV absorption proper-ties of bases (addition of sugar to the base slightly shifts the UV spectrum) it is possible again to judge about the selectivity of action on the bases of the same type using the difference spectra of mixtures of nucleosides and nucleotides. In the experiment we used the derivatives of uracil and adenine. Fig.3b shows difference UV spectra obtained after irradiation of acid aqueous solutions (pH = 2.2) of uridine, uridine - 5´-phosphate, adenosine, adenosine - 5´- phosphate. These spectra ar are seen to be similar to the difference spectra of uracil and adenine shown in Fig.3a.

Fig.5 shows the corresponding dependences of photoproducts yields versus irradiation intensity. From this figure one can make the following conclusion that the addition of sugar and then of phosphate to uracil sharply decreases the intensity of UV irradiation necessary for saturation of $S_0 \rightarrow S_1$ transition, that corresponds to the increase of the lifetime τ_1 of electro-nic state S_1. At the same time the addition of sugar and then of phosphate to adenine does not change essentially the life-time τ_1. Considering as before the UV pulse duration to be equal to 30ps we obtain τ_{1Urd} = 320βps, τ_{1pU} = 520βps, τ_{1Ado} = 41βps, τ_{1pA} = 90βps. These results are in agreement with the τ_1 estimates for nucleotides [10]. The quantum yield of two-step photoreactions changes insignificantly from bases to nucleosides and nucleotides. According to [7] we have φ_{Urd} = 2.8δ%, φ_{pU} = 3δ%, φ_{Ado} = 0.2δ%, φ_{pA} = 0.2δ%.

Note that the change in the lifetime τ_1 for Urd and pU leads to an increase of the selectivity of action at small

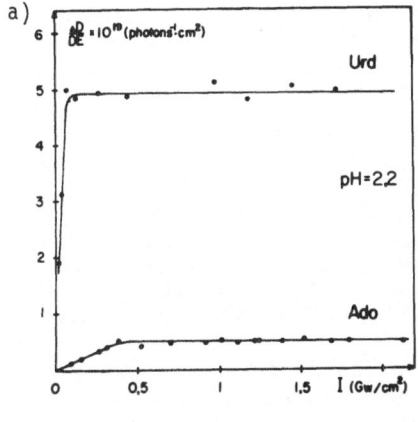

a)

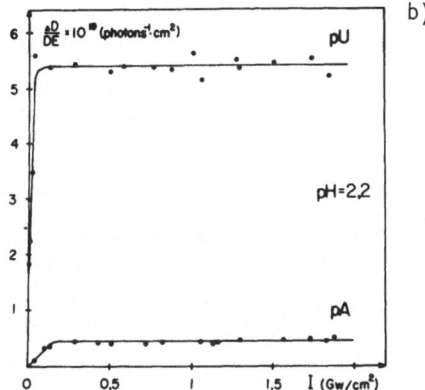

b)

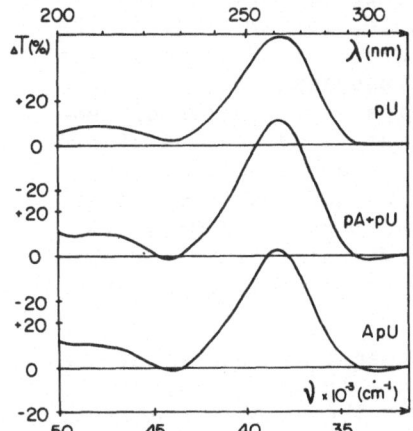

Fig. 5 The dependences of photoproducts yield on irradiation intensity (a) for uridine and adenosine (b) for uridine - 5´ - phosphate and adenosine - 5´ - phosphate. The irradiation wavelength λ = 266 nm

◄Fig. 6 Difference spectra obtained after UV irradiation of uridine - 5´ - phosphate, euqimolar mixture uridine - 5´ - phosphate - adenosine - 5´ - phosphate, dinucleotide adenylyl - 3´→5´ - uridine (pH = 2.2). The irradiation wavelength λ = 266 nm

irradiation intensities (for example for the pair uridine-adenosine S = 48 at I = 0.08Gw/cm²). At large irradiation intensities the selectivity of action on uracil S = 10 for the pair Urd - Ado and S = 12 for the pair pU - pA.

Finally we have carried out an experiment on selective action on uracil which is included in dinucleotide ApU (part of single-stranded RNA).

We have the following proofs of selectivity of action on uracil in the dinucleotide ApU at our disposal:

1) While the optical density change is small the difference spectrum of ApU is identical both to the difference spectrum of the mixture pU + pA and to the difference spectrum of pU.

2) Considerable change of the optical density leads to the fact that the difference spectrum of ApU remains similar only to the difference spectrum of the mixture pU + pA (Fig.6). In this case the maximum of the difference spectrum of ApU slightly shifts to the maximum of the difference spectrum of pA. Simultaneously in the difference spectrum of ApU there appear bands of pA photoproducts with λ = 227nm and λ = 296nm.

3) The UV absorption spectrum of the acid aqueous solution
of ApU after continuous irradiation is close to the UV spectrum
of pA.
Thus in this series of experiments we have proved the principal possibility of selective action on the bases of the same
type in native DNA and RNA by powerful picosecond UV laser
pulses. The nonlinear character of action is different for each
type of bases, that provides selectivity in spite of strong
overlapping in linear UV absorption spectra. It is essential
that the selectivity is preserved as nucleic acids fragments
containing bases irradiated become more complex. This provides
the possibility of practical realization of the methods of
selective laser photochemistry [2] on biological macromolecules.

Finally, let us underline once more, the term 'selectivity
of photochemical process' used in this work has a wider sense that it had in our previous work on laser-isotopically-se-
lective photochemistry (see, for example, review [1]). In the
particular case of isotopic species (atoms or molecules), the
selectivity was achieved by selective photoexcitation (single
or multi-step) only, because the chemical properties of isotopic species in initial or excited states, as well as of products of photochemical reactions, are almost identical for
different isotopes. In present work, the chemical properties
of different nucleic acid bases and their photoproducts are
not identical. It means that the observed selectivity of photochemical reactions can be caused by the selective two-step

excitation of different nucleic acid bases, as well as by the
difference of photochemical yields of highly-excited electronic states of different bases, even for the equal probability
of their two-step photoexcitation. One of the aims for our future experiments is to understand the origin of the observed
nucleic acid base selective two-photon photochemical reactions
induced by powerful UV picosecond laser pulses.

References

1. V.S.Letokhov, in " Progress in Optics ", vol.16, ed. by
   E.Wolf (North-Holland Publ. Co., Amsterdam - New York -
   Oxford, 1978), p.3.
2. V.S.Letokhov, in " Tunable Lasers and Applications ", ed.
   by A.Mooradian, T.Jaeger, P.Stokseth (Springer Series in
   Optical Sciences, vol.3, Springer-Verlag, Berlin - Heidelberg - New York, 1976), p.122.
3. S.L.Shapiro, A.J.Campillo, V.H.Kollman, W.B.Goad: Opt.Com-
   mun., 15, 308 (1975).
4. Yu.A.Matveetz, D.N.Nikogosyan, V.Kabelka, A.Piskarskas:
   Kvantovaya Elektronika (Russian), 5, 664 (1978).
5. P.G.Kryukov, Yu.A.Matveetz, D.N.Nikogosyan, A.V.Sharkov:
   Kvantovaya Elektronika (Russian), 5, 2348 (1978).
6. D.A.Angelov, G.G.Gurzadyan, D.N.Nikogosyan: Kvantovaya
   Elektronika (Russian), 6, N10 (1979).

7. P.G.Kryukov, V.S.Letokhov, D.N.Nikogosyan, A.V.Borodavkin,
    E.I.Budowsky, N.A.Simukova: Chem.Phys.Lett., 61, 375 (1979).
8. D.N.Nikogosyan, A.V.Borodavkin, N.A.Simukova, in " Spectro-
    scopy of Molecules, Crystals and Biological Systems ",
    (Estonian SSR Academy of Sciences, Tallinn, 1979), p.188.
9. G.J.Fisher, H.E.Johns, in " Photochemistry and Photobiology
    of Nucleic Acids ", vol.1, ed. by S.Y.Wang (Academic Press,
    New York, 1976), p.169.
10. A.A.Lamola, J.Eisinger: Biochim.Biophys.Acta, 240, 313 (1971).

Laser Selective Photobiology: Dye-Biomolecule Complexes

A. Andreoni, R. Cubeddu, S. De Silvestri, and O. Svelto

Centro di Elettronica Quantistica e Strumentazione Elettronica
Istituto di Fisica del Politecnico
Milano, Italy

G. Bottiroli

Centro di Studio per l'Istochimica, Università di Pavia
Pavia, Italy

1. Introduction

After the successful use of laser beams to perform selective photochemistry
[1], the possibility of laser selective photobiology has become of current
interest at a few laboratories. For this purpose, a first proposal, based
on two-step dissociation of a given biomolecule was made by Letokhov [2] .
In this proposal, however, level thermalization must be prevented in order
to retain selectivity. In the liquid phase, thermalization of the vibrational modes of a molecule usually occurs in a time of the order of a few
picoseconds, if not shorter [3]. Accordingly, this scheme requires picosecond
or even sub-picosecond pulses for both steps. Despite this difficulty, selective action on nucleic acid components have recently been reported by Kryukov
et al. [4]. The experiment was performed using high-intensity UV pulses.
A different proposal for selective laser photobiology was advanced by our
group in a previous work [5]. This proposal is based on two-step photoionization (or dissociation) of a dye molecule bound to a given biological
site. This process induces irreversible chemical damage to the biomolecule
[6] . The proposal makes use of the fact that the dye fluorescence lifetime
is a very sensitive function of the biological environment (i.e. of the binding site or of the biomolecule). Selectivity is achieved through the use of
two laser pulses with suitable delay. The advantages of such a scheme can
be summarized as follows: (i) Since specific stainings have been developed
for many biomolecules, the method is of somewhat general applicability.
(ii) Visible or near-UV photons are required for both steps. Selective photobiology can thus be performed "in vivo", since all biological materials
are transparent at these wavelengths.(iii) Nanosecond or sub-nanosecond, rather than picosecond pulses, with an intensity of a few MW/cm^2, are required.
    This work presents a detailed discussion of the various selectivity proc-
esses. In particular, it will be shown that, in some cases, two nanosecond
pulses with zero delay (which results in an even simpler configuration) suf-
fice to perform the selective process. A few experimental results, which confirm some of these ideas, are also presented.

2. Schemes of Selective Action

We will consider a suitable fluorescent dye bound to a given biomolecule. We
will assume that, depending on the binding site, two complexes (henceforth
called 1 and 2) with different fluorescence lifetimes τ_1 and τ_2 ($\tau_1 < \tau_2$) are
formed. Selective action on the two complexes is performed by two-step ionization (or dissociation) of the dye by two laser pulses of suitable duration
τ_p and suitable delay τ_d. Under these conditions a few selective schemes can
be considered. To this end Fig. 1 shows the electronic states of the two com-

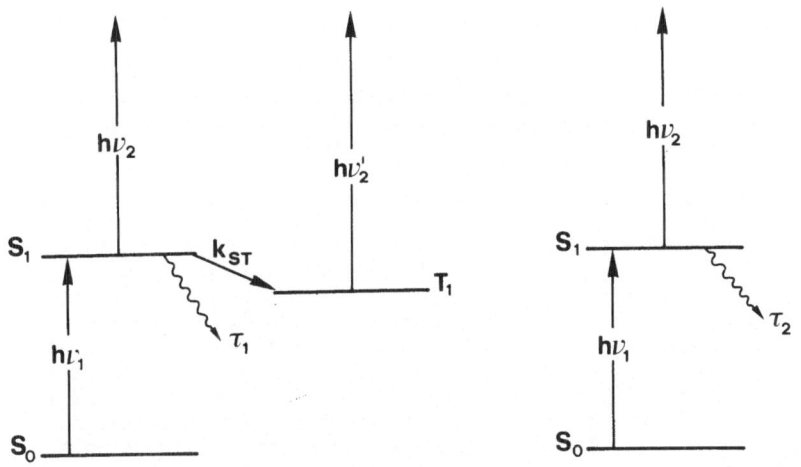

Fig.1 Electronic states of complexes 1 (left) and 2 (right)

plexes considered in this work.

In the first case, τ_p is assumed to be appreciably smaller than both τ_1 and τ_2, and τ_d is such that $\tau_1 < \tau_d < \tau_2$. When the second pulse interacts with the dye, complex 1, due to its faster decay, will show a S_1-state population appreciably smaller than that of complex 2. Complex 2 will therefore be preferentially ionized (or dissociated). If the two-step process only arose through the S_1 state, the selectivity of the process would follow the expression $\exp(-\tau_d/\tau_2)\exp(-\tau_d/\tau_1)$: i.e., it would increase by increasing τ_d. However, it is also necessary to take into account the possibility of a two-step process through the dye triplet state T_1. Since $\tau_1 < \tau_2$, the intersystem rate k_{ST}, and hence the T_1 population of complex 1, is larger than that of complex 2. To retain good selectivity, the ionization cross-section starting from S_1-state (σ_s) has to be appreciably larger than that starting from T_1-state (σ_T). Although a detailed calculation is beyond the scope of this work, we merely wish to point out that if $\tau_1 \ll \tau_d \ll \tau_2$, complex 1 will be found on T_1, while complex 2 will still be on S_1. The selectivity will thus be limited by σ_s/σ_T.

In the second scheme, τ_p is assumed to be of the order of (or larger than) τ_2, and τ_d is set at zero. The S_1 state populations of both complexes will reach a steady-state condition wherein the S_1 population of complex 2 is appreciably larger than that of complex 1. Again, complex 2 will be preferentially ionized (or dissociated). In this case, the selectivity is smaller than that of the previous case. This scheme appears to be simpler, however, and it also allows action, even when the frequencies of the two pulses need to be the same. Note also that, within the limit $\tau_1 \ll \tau_2$, the selectivity is again of the order of σ_s/σ_T.

So far, schemes for selective action on complex 2 have been considered. To perform selective action on complex 1, we choose τ_d in such a way that $\tau_2 < \tau_d < \tau_T$, in which τ_T is the lifetime of T_1 (usually much larger than τ_2). In this case, when the second pulse arrives, complexes 1 and 2 will be found both on T_1; but, since $\tau_1 < \tau_2$, the population of complex 1 will be larger than that of complex 2. Complex 1 will now be preferentially ionized (or dissociated.

3. Two-step Selective Action on Acridine Dyes

To test some of the ideas discussed in the previous section, a few experiments have been performed on acridine dyes, either in a buffer solution or bound to synthetic polynucleotides. When an acridine dye is bound to the DNA, the dye lifetime τ strongly depends on the DNA base-pair sequences where the dye is intercalated [5]. Thus, we have: (i) $\tau = \tau_2 = 10 - 20$ ns when the dye intercalates is AT-AT base-pair sequences. (ii) $\tau = \tau_1$ 0.5-1 ns when the dye intercalates in either a GC-AT or a GC-GC base-pair sequence.

In an initial experiment, a 3.10^{-6}M solution of proflavine (PF) in an 0.2 M Acetate buffer (pH = 4.6) was used. The experimental arrangement is shown in Fig. 2. The first excitation pulse (λ_1 = 430 nm) of $\sim$ 250 ps duration, $\sim$ 100 kW peak power, and repetition rate up to 100 Hz was provided by a dye laser specially designed by us [7] . The dye laser was pumped by an atmospheric-pressure N_2 laser, which generates pulses of $\sim$ 500 ps duration and $\sim$100 kW peak power [8] . The second excitation pulse at λ_2= 337.1 nm was provided by the other output of the N_2 laser. This pulse was delayed by a suitable optical delay line. The two beams, focused into the sample cell, were co-linear and in opposite directions.

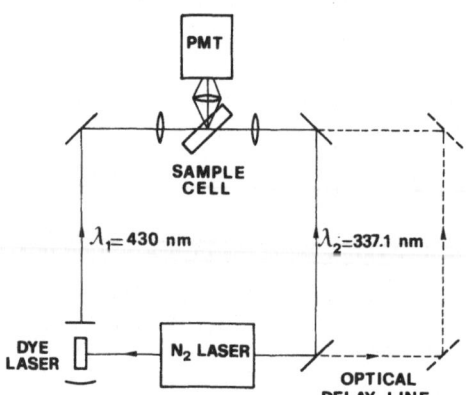

Fig.2 Experimental arrangement

The first measurements were performed at zero delay between the two beams in order to demonstrate the biphotonic character of the process. It was found that: (i) When both beams were present, rather strong decomposition of PF dye occurred. This decomposition was monitored through the reduction of the peak of the fluorescence signal, as observed with an XP 1020 photomultiplier. (ii) No appreciable decomposition was observed, within the same interaction time, when one of the two beams was blocked. These results show that the dye decomposition observed occurs by biphotonic process.

A second set of measurements was performed to show that the idea of using two suitably delayed beams was actually working in our case. For this purpose the N_2 laser beam was delayed before entering the sample cell. The delayed pulse thus finds a smaller population in S_1 and a correspondingly larger population in T_1 as compared with the undelayed case. If we assume a larger dissociation cross-section starting from S_1 than from T_1 , we would therefore expect a reduction in dissociation rate upon increasing the delay between the two pulses. This is, indeed, what is found experimentally, as shown in Fig. 3. In this figure the experimental values of the fractional reduction in dye cell fluorescence (i.e. the fraction of total dye decomposed), at three different delays between the two beams are indicated as square dots. The irradiation parameters were kept the same for the three cases. The

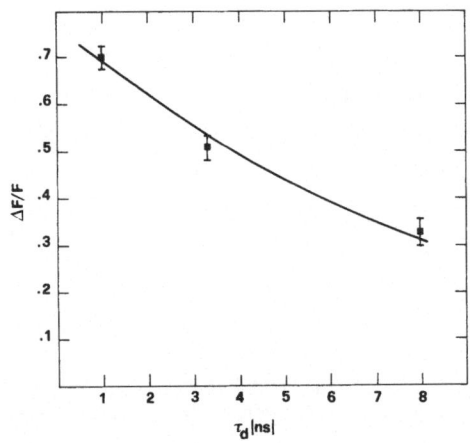

Fig.3 Fractional reduction of fluorescence $\Delta F/F$ as function of time delay τ_d.

solid curve is a theoretical best fit with our results, assuming a dye life-time of 4.3 ns (as measured by us) and a dye fluorescence quantum yield of 0.7 [9] . The calculation takes into account the dissociation of the dye from both S_1 and T_1 and the reduction of the dye concentration during the irradiation time [10] . The value of the ratio σ_S/σ_T turns out to be ~ 2.7.

Another experiment was performed on the acridine dye Quinacrine Mustard (QM). In this case, the same N_2 laser pulse was able both to excite and to disso-ciate the dye. Accordingly, a different N_2 laser (5 ns pulse duration, 200 kW peak power, and repetition rate up to 50 Hz) was used. The amount of dye de-composition was again monitored by measuring the fluorescence intensity of the dye through a photomultiplier. An initial set of measurements was per-formed on QM ($3 \cdot 10^{-6}$M) in an 0.2 M Acetate buffer solution (pH = 4.6) to show the biphotonic character of our photodissociation. To this end we define a damage probability p per laser pulse as the quantity p = $\Delta N/N$, in which N = N(n) is the dye concentration after the n-th laser pulse and ΔN is the reduction of this concentration after the next laser shot. The experimental results show that, for light intensity I smaller than a few MW/cm^2, the dam-age probability p is proportional to I^2, which is indicative of a biphotonic process. At higher intensities, saturation of $S_0 \rightarrow S_1$ transition sets in [11].

Having established the biphotonic character of our interaction, a second set of measurements was performed by irradiating the QM dye bound to either Poly dA - Poly dT or to Poly dG - Poly dC synthetic polynucleotides in the same buffer solution. A dye concentration of 10^{-6}M and a DNA phosphorus con-centration of 10^{-4}M were used. The idea of the experiment is selectively to dissociate the dye bound to Poly dA - Poly dT (complex 2, see previous sect-ion) as compared to that bound to Poly dG - Poly dC (complex 1). From the observed fluorescence decrease of complex 1 and 2 irradiated under the same conditions, and at several irradiation times, we obtain the damage probabi-lities p_1 and p_2 for the two complexes. Our experimental results give $p_2/p_1 \simeq 4.7$. Quite good selective action on complex 2 has thus been demon-strated.

It is worth pointing out that our biphotonic ionization (or dissociation) channels, unlike the usual photodynamic action does, not require oxygen [11] .

4. Conclusions

In this work, a few ideas on laser selective two-step dissociation of dye molecules attached to specific environment have been discussed. Selectivity is achieved by using two-laser pulses of suitable duration and delay to perform the two-step process. Some of these ideas were tested in preliminary simple experiments on the acridine dyes, Proflavine and Quinacrine Mustard. In particular, a larger damage probability has been shown to occur when the QM molecule is bound to Poly dA - Poly dT, compared with when it is bound to Poly dG - Poly dC. The ratio of the two probabilities (i.e. the selectivity of the process) is, in our case, 4.7.

In conclusion, we may say that our proposed schemes of dissociation seem to provide for simple and somewhat general means of laser-selective photobiology.

References

[1] V.S.Letokhov and C.Bradley Moore, in: Chemical and Biochemical Applications of Lasers, Vol. 3, ed. C.Bradley Moore (Academic Press Inc., New York, 1977) p. 1
[2] V.S.Letokhov, in: Tunable Lasers and Applications, eds. A.Mooradian, T. Jaeger and P. Stokseth (Springer-Verlag, Berlin, Heidelberg, New York, 1976) p. 122
[3] A.Lauberau and W.Kaiser, in: Chemical and Biochemical Applications of Lasers, Vol. 2, ed. C.Bradley Moore (Academic Press Inc., New York, 1977) p. 87
[4] P.G. Kryukov, V.S.Letokhov, D.N.Nikogosyan, A.V.Borodavkin, E.I.Budowsky and N.A. Simukova, Chem. Phys. Lett. 61, 375 (1979)
[5] A.Andreoni, C.A.Sacchi and O.Svelto, in: Chemical Biochemical Applications of Lasers, Vol. 4, ed. C.Bradley Moore (Academic Press Inc., New York, 1979) p. 1
[6] A.Gräslund, A.Rupprecht and G.Strom, Photochem. Photobiol. 21, 153 (1975)
[7] R.Cubeddu, S.De Silvestri and O. Svelto, unpublished
[8] R.Cubeddu and S.De Silvestri, Opt. Quant. Electr. 11, 276 (1979)
[9] R.Rigler, in: Chromosome Identification. Nobel Symposium XXIII, eds. T. Caspersson and L. Zech (Academic Press Inc., New York, 1973) p. 335
[10] A.Andreoni, R.Cubeddu, S.De Silvestri, P.Laporta and O.Svelto, unpublished
[11] A.Andreoni, R.Cubeddu, S.De Silvestri and P.Laporta, unpublished

Picosecond Fluorometry of the Exciton Diffusion in Green Plant Antenna Chorophyll

L.B. Rubin[1], V.Z. Paschenko[2], and A.B. Rubin[2]

[1] Faculty of Physics, Moscow State University,

[2] Faculty of Biology, Moscow State University, Moscow, 117234, USSR

1. Introduction

In the past few years, due to the continued aggravation of the energy crisis, much scientific activity has been directed towards the search for new alternative energy sources. In this regard, there is heightened interest in sunlight as a source of energy, and the most valuable prospects for practical solar energy conversion might be based on the photosynthetic model, since photosynthesis is an extremely effective means whereby nature captures solar energy, converts it into separated charges, and stores it in useful chemical form.

In contrast with other bioenergetic processes, the specificity of the operation of the photosynthetic apparatus lies in the very primary events that occur in the picosecond (10^{-10} s) time domain and lead to charge separation . Studies of such ultrafast photoreactions have become a reality only since the development of a new optical method - picosecond spectroscopy.

It is now well established that in photosynthesis the primary processes of excitation energy conversion take place in the following several stages. The excitation quanta of light are absorbed by aggregates of pigments, called photosynthetic units (PSU). The PSU consists of several hundreds of pigments, mainly chlorophyll (Chl), and carotenoids, which together act as a light-harvesting-antenna (LHA) system. The role of the antenna is to transfer energy to a specialized trap or reaction center (RC). This process takes less than 1 ns. Charge separation and stabilization of the separated charges than occurs within the RC in about 10^{-10} s. It is now universally recognised that PSUs form close associations, capable of exchanging energy with one another (the multitrap, or "lake" model).

In green plants, photosynthesis is produced by two different light reactions, sensitized by Photosystems I and II [1]. As a consequence of two photosystems, the pigment complex in green plants has a complicated organization, and its LHA channels the energy either to the RC of PSII (RCII) or to a focusing antenna of PSI, to be then trapped by its RC (RCI) [2,3]. The two photosystems have different spectral properties: PSI fluoresces at 735 nm and the fluorescence maximum of PSII at 685 nm (77°K). The fluorescence of PSII is believed to be due to the LHA Chls; the PSI fluorescence is ascribed to the long wavelength forms of Chls in its own antenna system [4] .

The overall quantum efficiency of the primary photosynthetic events is very high - about 0.8-0.9 for energy migration from the LHA to the RC [5]

and 0.97-0.99 for charge separation [6] . Obviously, knowledge on the physical mechanisms that present such unique characteristics might help much
in attempts to create artificial converters of solar energy.

The advent of high resolution picosecond fluoremeters have made possible
measurements, with a resolution of 10^{-11} s, of fluorescence kinetics in higher plant chloroplasts and subchloroplast fragments. First, picosecond
fluorometric observations were carried out by high-intensity picosecond
pulse train excitation, the pulse energies used being as high as $5 \cdot 10^{14}$
photons$\cdot$cm^{-2}, or even more. Subsequent studies have demonstrated that under
high-intensity excitation conditions nonlinear decays of fluorescence take
place with a dramatic drop in their quantum efficiency (ϕ) and lifetime (τ)
[7,8,9] . The phenomena was explained in terms of bimolecular singlet-singlet and singlet-triplet annihilation processes [10-12] .

The significance of this work is not so much that it provides true estimates for $\tau_{f\ell}$, as in an impetus it has given to singlet diffusion studies
in the photosynthetic pigment apparatus. This means that studies of the effects of excitation intensities on fluorescence lifetime and quantum efficiency may provide information about coefficient (D) and diffusion distance
(L) for excitons formed within the LHA [8] . The creation and destruction
of excitons, n (t), in the case of $S_I + S_I$ annihilation, when the singlets
come within their interaction radius, can be adequately described by a standard kinetic equation:

$$\frac{dn(t)}{dt} = C(t) - \beta n(t) - \frac{1}{2} \gamma_{ss} n^2 (t) \qquad (1)$$

where C(t) is a function of the singlet excitation source; $\beta = 1/\tau$ is a deactivation rate constant for low-intensity excitation; γ_{ss} is a constant for
an $S_I + S_I$ annihilation process. It is necessary to mention that application eq. (1) is correct only on some assumptions [11] . From eq. (1) the excitation intensity dependences of fluorescence lifetime, $\tau_{1/e}$, and its
quantum efficiency ϕ/ϕ_0 can be found:

$$\tau_{1/e} = \frac{1}{\beta} \ln \frac{e + \frac{\gamma_{ss} n(0)}{2\beta}}{1 + \frac{\gamma_{ss} n(0)}{2\beta}} \qquad (2)$$

$$\phi/\phi_0 = \frac{2\beta}{\gamma_{ss} n(0)} \ln \left[1 + \frac{\gamma_{ss} n(0)}{2\beta} \right] \qquad (3)$$

Using eq. (2) and (3), we can obtain a family of theoretical curves -
$\tau_{1/e}$ and ϕ/ϕ_0 versus excitation intensity for different values of γ_{ss} . Comparison of such curves with experimental results will make it possible to
estimate the value of γ_{ss} in a certain system of investigation [8,11] .

In terms of a structural model for the LHA complex in green plants, it
is now generally accepted that excitation energy absorbed in the antenna
migrates to the RCs of PSII and PSI [3] . It is believed that excitation

trapping at the RCs is a dominant process that accounts for short lifetimes of electronic excitations in the LHA. As the theory predicts, RCs can be regarded as quenchers. Obviously, for the exciton diffusion mechanisms to be understood, we need experiments with material free of those quenchers. In fluorescence investigations carried out to this end, we studied excitation intensity effects on τ and $\phi/\phi\circ$ in pea chloroplasts and their fragments.

2. Methods

A streak camera has the advantage of permitting the continuous recording of fluorescence decay excited by a single laser pulse [13,15] . An instrumental setup employing a streak camera detecting system is described earlier [16] . A second harmonic 530 nm pulse from Nd: glass laser is split into a reference and excitation beam. The reference one goes directly to the streak camera to monitor the shape and energy of the laser excitation pulse. Fluorescent light, after going through the spectrograph, is focused onto a photocathode in the camera. Photoelectrons, produced in proportion to the light intensity, are accelerated in an electric field and then deflected by a voltage ramp onto a phosphorescent screen so that electrons released from the cathode at different times hit the screen at different locations. This provides a means of linearly calibrating the screen in time. Because of very fast sweeps, streak cameras have a time resolution of the order of 10^{-12} s, while their sensitivity is not worse than that of photomultipliers [17,18] . The phosphorescence can be photographed directly on film, or detected on a silicon vidicon screen, and the image can be stored in an OMA. After data processing, the output represents the fluorescence decay and the laser excitation pulse as a ratio of the area under the fluorescent decay curve to that under the laser pulse (duly calibrated in units of energy).

3. Results

In the first series of experiments, we used RC-free LHA preparations[19] . The lifetime of fluorescence 685 nm was found to be reduced from 3000 ps to 150 ps, as the excitation intensity was varied from $3\cdot10^{12}$ to 10^{17} photons· cm^{-2} (see Fig.1). Non exponential decays were observed at high intensities and a mono exponential one at low excitation intensities. Experiments with entire chloroplasts and PSII-enriched chloroplast particles also showed a dramatic changes in τ and ϕ/ϕ_0 at 685 nm with increase in the intensity (refer to Fig. 2,3).

According to Duescence's well-known hypothesis, RCs act as quenchers for antenna excitation only when in the reduced ("open") state. One would expect, considering the high quantum efficiency of photosynthesis, that the oxidation of the RC and its transition into the "closed" state would be attended by a rise in the lifetime and quantum efficiency of LHA fluorescence to values that are normally encountered in Chl solutions. Meanwhile, our experiments showed only a 2-3 fold rise in τ_{fl} for PSI and PSII, with the RC driven to the "closed" state [13,20] . Bearing this in mind, we carried out a series of experiments with entire chloroplasts and PSII subchloroplast particles containing only closed RCs (see Figs. 4 and 5).

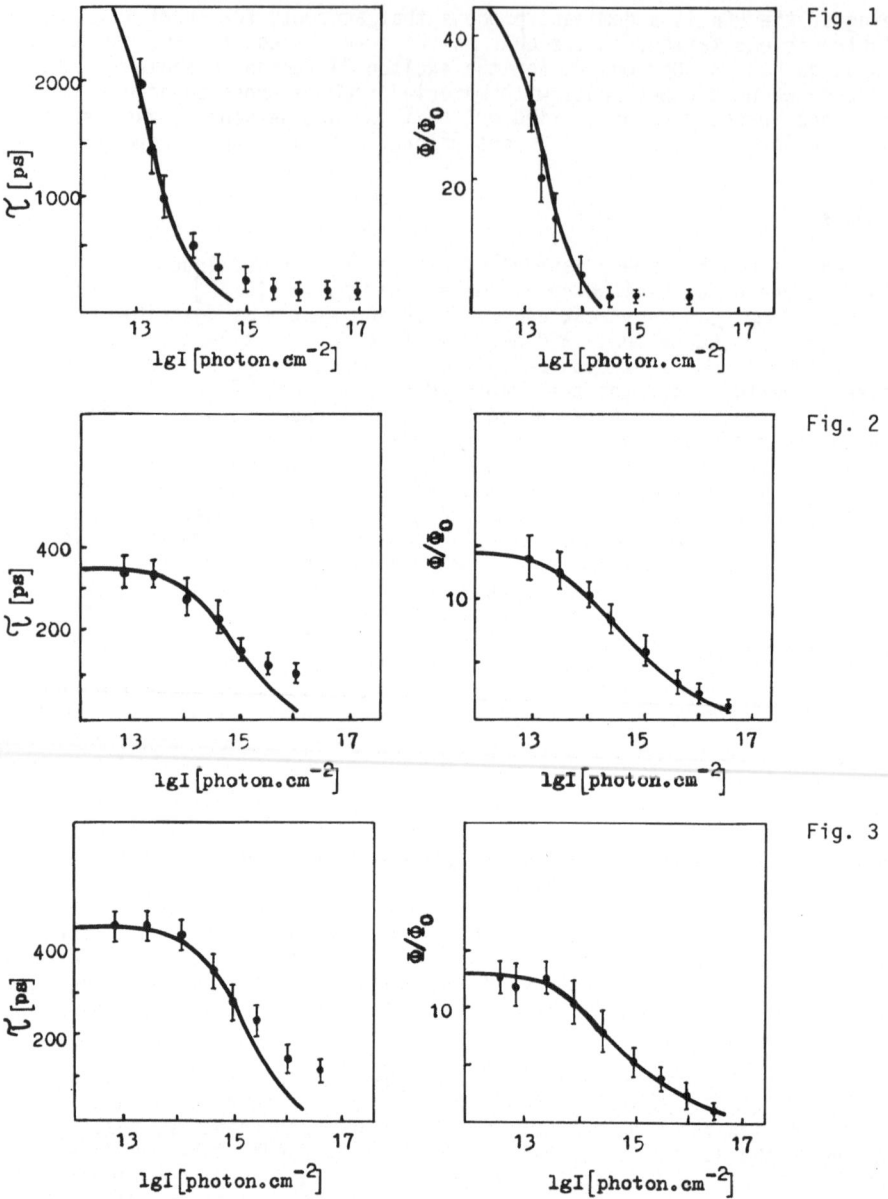

Fig. 1

Fig. 2

Fig. 3

Figs. 1-3. Figure captions see opposite page.

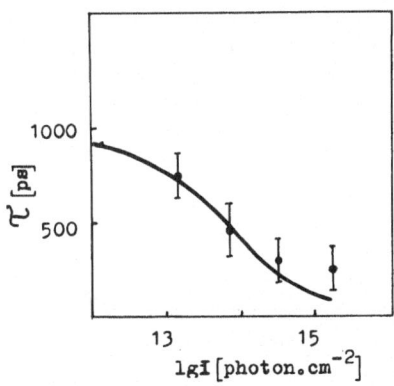

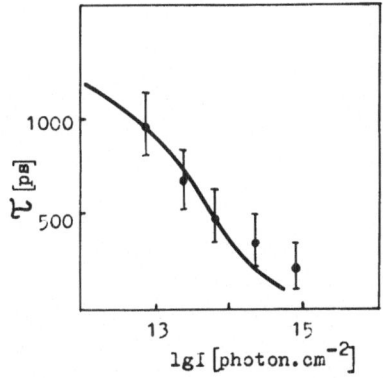

Fig. 4. Dependence of entire chloroplast fluorescence (RCII "closed")
on the energy of 350 nm excitation.
————Theoreical curve for $\gamma = 4 \cdot 10^{-8}$
$cm^3 s^{-1}$

Fig. 5. Dependence of PSII particle
cluorescence (RCII-"closed") on the
energy of the 350 nm excitation.
————Theoretical curve for γ
$7 \cdot 10^{-8} cm^3 s^{-1}$

Examination of the data in Figs. 2-5 shows, first of all, that τ_{fl} (LHA)=
3000 ps is close to those reported for Chl solutions [13,21] , thus suggesting that there is no concentration effect on Chl in vivo fluorescence
although, in solution at high Chl concentration, there occurs strong
fluorescence quenching.   In fact, in a Chl b chlorophorm solution the
fluorescence lifetime of 572 ps was reduced to 15 ps when the concentration
was increased from 3.8 mM to 0.1 M [15] .

It is probable that the lack of concentration quenching of Chl in vivo
is the consequence of its specific imbedding and the way it interacts with
proteins of the LHA. If this conclusion is true, it is evident that Chl-
protein complexes in vivo play an extremely important role, since it is
those very structures that provide the unique property of the LHA: efficient light harvesting and transfer of the energy to the RC with no losses
associated with concentration quenching [22-24] .

Variations in lifetime observed in our preparations are obviously due to
different quenching activities of RCII's and RCI's. In fact, in samples
containing only RCII's (PSII-enriched chloroplast fragments) the lifetime
of antenna pigment fluorescence (normally 3000 ps) is reduced to 1200 ps,
when the RCII's were driven to the closed state and to 450 ps when they

---

◄Fig.1. Dependence of the lifetime (τ) and the relative quantum yeld (ϕ/ϕ_0)
of the fluorescence of the LHA from pea chloroplasts on the energy of the
530 nm excitation pulse.
—— Theoretical curves for $\gamma = 7 \cdot 10^{-8}$ $cm^3 \cdot s^{-1}$

◄Fig.2. Dependence of the τ and ϕ/ϕ_0 of the entire chloroplasts fluorescence on the energy of the 530 nm excitation pulse —— Theoretical curves
for $\gamma = 1,5 \cdot 10^{-8}$ $cm^3 \cdot s^{-1}$

◄Fig. 3. The values of the τ and ϕ/ϕ_0 of the PSII particle as a function of
the 530 nm excitation pulse energy. ——— Theoretical curves for $\gamma = 0,7 \cdot 10^{-8}$
$cm^3 \cdot s^{-1}$

were opened. When both RCI and RCII quenchers were present (entire chloroplasts) the reduction in τ of the LHA fluorescence was even more pronounced:
down to 900 ps with closed RCII's, and to 340 ps with open RCs. From these
data, the rate constants for energy migration from the LHA to the RCs with
their nearest pigments surroundings K_M^{II} and K_M^I, can be calculated:

$$K_M^{II} = \frac{1}{\tau(PS\ II)} - \frac{1}{\tau(LHA)} \tag{4}$$

$$K_M^I = \frac{1}{\tau(Chloroplast)} - \frac{1}{\tau(LHA)} - K_M^{II}$$

where $\tau PSII$, τLHA, and $\tau(Chloroplast)$ are fluorescence lifetimes for PSII,
LHA and entire chloroplast preparations. Calculation give: $K_M^{II} = 1.82 \cdot 10^9$
s^{-1} (open RCsII);
$K_M^I = 0.64 \cdot 10^9$ s^{-1}. Hence, for normal conditions with open RCIIs $K_M^{II}/K_M^I = 3$,
i.e. RCII trap 3 times as much energy as RCI's. The data show only mean
excitation times for the LHA. Further information on exciton dynamics occurring within the LHA can be obtained by mathematical processing. By comparing theoretical and experimental curves (see Figs. 1-5) we found the
values of γ_{SS} for singlet-singlet annihilation in the presence of RCII and
RCI quenching centers. In the case of three-dimensional isotropic diffu-
sions, there is the following relationship between γ_{SS} diffusion coefficient D [25] :

$$\gamma_{ss} = 8\pi DR \tag{5}$$

R, here, is the radius of the interaction between the two excitons with an
efficiency of P~I. The validity condition for Eq. (5) is:

$$L = \sqrt{\tau D} >> R \tag{6}$$

(L= the diffusion distance). These conditions can be satisfied only at mo-
derate excitation levels, when excitation diffusion can take place. Assuming that the LHA has a rigid three-dimensional crystal configuration with
Chl molecules in its cross-joints, the crystal constant for a Chl concentration of 0.I M will be equal to 20A°. In all likelihood, there is strong
exciton-phonon coupling in the LHA complex, and the excitation, once localized on one of its Chl molecules, moves further by jumps [23] . The probability of excitation transfer diminishes with the sixth power of the distance, r^6, over which it is transferred. Therefore, as a first approxima-
tion, exciton jumps over distances greater than the crystal constant may
be neglected. In fact, this means that the effective radius for exciton-
exciton interactions can be taken as a crystal constant [26,27] i.e. 20A°.
As theory predicts, the probability, W, of a non-coherent exciton jump is
a function of a diffusion coefficient, D [28] :

$$W = D \cdot R^2 \tag{7}$$

Using the values of γ_{ss} given in Figs. 2-5, and taking R as 20A°, Eqs. (5-
7) can be solved to obtain D, L, and W (Table 1). Note that calculations
were made on the assumption that the annihilation probability, P, is unity,
while in reality P<I. Therefore, the estimates in Table 1 correspond to
their lowest probable limits.

Our value for the antenna diffusion constant ($D \sim 2 \cdot 10^{-2}$ $cm^2 s^{-1}$) is somewhat greater than those reported in the literature [10,11,19]. We believe it to be due to the material used. No exciton diffusion experiments with RC-free LHA complex preparations have been performed as yet. Obviously in the presence of RC quenchers, exciton diffusion is not homogeneous. Moreover, after excitation, some of RCII quenchers become closed, a fact that has been disregarded thus far. The essential feature is a rather large path $L \sim 900\text{A}$, for excitation diffusion in the LHA complex, which is taken as evidence for a multitrap domain model for PSU organization.

It is known that lowering of the temperature leads to an increase in fluorescence lifetime. The fluorescence activation energy, ΔE, was found by us to be 0.02 eV. [30]. This energy is in fact a measure of the energy depth of an RCII quencher with respect to the energy level of the localized excitation.

Table 1   Parameter of the exciton diffusion in green plant photosynthetic apparatus

Sample	$\gamma_{ss} \cdot 10^8$ $cm^3 s^{-1}$	D $cm^2 s^{-1}$	L $\text{A}°$	W s^{-1}	reference
* LHA from pea Chloroplasts	7 ± 3	$2 \cdot 10^{-2}$	900	10^{12}	[31,32]
* PSII particles (RCII-open)	1 ± 5	$2,5 \cdot 10^{-3}$	110	10^{11}	[31,32]
* PSII particles (RCII-closed)	7 ± 3	$2 \cdot 10^{-2}$	490	10^{12}	[31,32]
**Spinach Chloroplasts	0,5	10^{-3}	200	$3 \cdot 10^{10}$	[11,29]

(*) three-dimensional model; (**) two-dimensional model

Taking the probability, $\vec{W}$, for an excitation jump from an antenna Chl molecule to the RCII as being equal to 10^{12} s^{-1} (Table 1) and using the ΔE value of 0.02 eV, the probability of a reverse excitation jump can be extimated as

$$\overleftarrow{W} = \vec{W}e - \frac{\Delta E}{kT} = 0.45\vec{W} \tag{8}$$

Thus, the probability of an exciton's leaving the RCII is as much as 0.45 W. The long excitation lifetime (3000 ps) and the wide diffusion path ($L \sim 900 \text{A}°$) are probably factors that, together, provide optimal conditions for a multiple quenching at the RC and, consequently, the high-quantum operation of photosynthesis.

REFERENCES`

1. Emerson, R, 1958, Ann.Rev.Plant Physiol. 9: 1-24.
2. Butler, W.L., and Kitajima, M., Biochimica et Biophysica Acta 396: 72-85.
3. Butler, W.L., 1978, Ann.Rev.Physiol. 39: 345-378.

4. Saton, K., and Butler, W.L. 1978, Biochimica et Biophysica Acta 502: 103-110.
5. Wraight, C.A., and Clayton, R.K., 1974, Biochimica et Biophysica Acta 333: 246-260.
6. Campillo, A.J., and Shapiro, S.L., 1978, Photochem. Photobiol. 28: 975-989.
7. Mauzerall, D., 1976, Biophys. J. 16: 87-92.
8. Campillo, A.J., Shapiro, S.L., Kollman, V.H., Winn, K.R., and Hyer, R. C., 1976, Biophysical J. 16: 93-98.
9. Campillo, A.J., Kollman, V.H., and Shapiro, S.L., 1976, Science 193: 227-229.
10. Swenberg, C.E., Geacintov, N.E., and Pope, M., 1976, Biophysical J. 16: 1447-1452.
11. Geacintov, N.E., Breton, J., Swenberg, C.E., and Paillotin, G., 1977, Photochem.Photobiol. 26: 629-638.
12. Swenberg, C.E., Geacintov, N.E., and Breton, J., 1978, Photochem. Photobiol. 28: 999-1006.
13. Paschenko, V.Z., Rubin, L.B., Rubin, A.B., Tusov, V.B., and Frolov, V. A., 1975, J.Tech.Phys. 45: 1122-1127 (Russian).
14. Paschenko, V.Z., Protasov, S.P., Rubin, A.B., Timoveev, K., Zamanova, L. M., and Rubin, L.B., 1975, Biochimica et Biophysica Acta 408: 147-153.
15. Shapiro, S.L., Kollman, V.H., and Campillo, A.J., 1975, FEBS Letters N 3,54, 358-362.
16. Rubin, L.B., and Rubin, A.B., 1978, Biophysical J., 24,1: 84-92.
17. Bradley, D.J., and Sibbett, W., 1975, Applied Physics Letters 27,7: 382-384.
18. Babenko, V.A., Kudinova, T.A., Malishev, V.I., Prohorov, A.M., Cichev, A.A., Tolmachev, A.U., and Shelev, M.Ja., 1977, Geft Letters 25,8: 306-369 (Russian).
19. Guljaev, B.A., and Tetenkin, B., 1979, Dokl. AN USSR 248,3: 345-347.
20. Paschenko, V.Z., Kononenko, A.A., Protasov, S.P., Rubin, A.B., Uspenskaya, N.Ya., and Rubin, L.B., 1977, Biochimica et Biophysica Acta 461: 403-412.
21. Brody, S.S., and Rabinowitch, E., 1957, Science 125: 555.
22. Lutz, M., 1977, Biochimica et Biophysica Acta 460: 408-430.
23. Paillotin, G., 1977. Proceeding of the Fourth International Congress on Photosynthesis, U.K., 4-9 September, 1977. Editors: Hall, D.O., Coombs, J., Goodwin, J.W. London, The Biochemical Society, London and Colchester, 33-44.
24. Korvatovsky, B.N., Tusov, V.B., Paschenko, V.Z., Rubin, L.B., Rubin, A.B., and Guliaev, B.A., 1979, Dokl.Acad.Nauk. USSR 247, 978-982.
25. Suna, A., Phys. Review B. 1 N 4: 1716-1739.
26. Agranovich, V.M., 1968, A Theory of excitons, Moscow, Nauka (Russian).
27. Agranovich, V.M., and Galaninm, M.D., 1978, Transfer of the energy of electronic excitation in condensed media, Moscow, Nauka (Russian).
28. Trifai, M., 1956, Czechoslovat Journal of Physica, 6, N, 6: 533-550.
29. Geacintov, N.E., Breton, J., Swenberg, C., Campillo, A.J., and Shapiro, S.L., 1977, Biochimica et Biophysica Acta, 461, 306-312.
30. Guliaev, B.A., Tetenkin, V.L., Pomerantseva, O.M., 1979, Dokl. Akad. Nauk USSR, 248, 752-755.
31. Paschenko, V.Z., Rubin, L.B., and Rubin, A.B., 1979, 3rd Conference on Luminescence, Conference Digest, vol. 1, Szeged, Hungary, September 4-7. 193-196.
32. Rubin, L.B., Korvatovsky, B.N., Braginskaja, O.V., Paschenko, V.Z., Paersche, H., Tusov, V.B., 198°. Molec. Biology (Russian, in press).

Subpicosecond Photodissociation and Time Resolved Spectroscopy of Recombination Processes in Carbonmonoxyhemoglobin

J.L. Martin, R. Astier, A. Migus, and A. Antonetti

Laboratoire d'Optique Appliquée, Ecole Polytechnique
Ecole Nationale Supérieure de Techniques Avancées, F-91120 Palaiseau, France

1. Introduction

The development of short laser pulses in the last decade provided new tools for spectroscopic studies of phenomena taking place on a picosecond time scale.

The extension of this technique in the subpicosecond regime offers new exciting possibilities by probing the previously unsolved elementary steps of biological or chemical reactions.

In some recent experiments, we have explored the early time spectral transients following photodissociation of carbonmonoxyhemoglobin.

The photolysis of hemoglobin complexes has been previously performed by SHANK et al. [1] in the subpicosecond time scale. They used a single sub-picosecond pulse (0.5 psec) at 615 nm selected from a CW mode-locked dye laser to excite the samples, and a much weaker delayed image of the pulse is used to probe the induced absorption as a function of time-delay. The repetition rate was 10 kHz and the energy per pulse was a few nanojoules. Under these conditions, it was concluded that CO dissociation occurs in less than 1 psec, but that no photolysis of HbO_2 occurred. The results for HbO_2 seem to be a consequence of a low quantum yield for dissociation of HbO_2 (~ 0.008).

More recently, GREENE et al. [2] investigated transient absorption spectra in the soret and visible regions after excitation of HbO_2 and HbCO samples with 353 or 530 nm pulses, 10 psec duration. They found transient species that appeared within 8 psec and exhibited considerably broadened deoxyhemoglobin-like spectra, which persisted during 680 psec.

A major aim of the present work was to measure transient spectra in the visible region with subpicosecond resolution. To reach this goal, subpico-second pulses from a passively mode-locked CW dye laser have been amplified to produce pulses with a peak intensity of 2 GW while maintaining a 0.7 pico-second pulse width. These amplified pulses have been used to generate continua with the same duration, extending from 0.30 μm to 1.2 μm.

We describe in this paper preliminary results obtained by photolysis of carbonmonoxyhemoglobin with the amplified pulses at 615 nm and probing the initial photoproduct in the 450 nm – 500 nm region.

2. Materials and Methods

Preparation of materials : Human Hb was obtained by standard methods [3]. Dilution to the final concentration were reached by using of a 0.15 M potassium phosphate buffer (pH 7.35). The solutions were degassed by pumping in a glass bulb attached to the sample cell. The HbCO was formed by admission of CO at 1 atm above the solution. The experiments were performed at room temperature.

Subpicosecond pulse generation : for these spectroscopic applications, it is desirable to amplify low energetic subpicosecond pulses [1] to higher powers. Our arrangement is shown schematically in fig. 1. The input pulse is obtained from a passively mode-locked CW dye laser [4] which utilized a prism (P) as dispersive element and two free flowing dye streams. The Rhodamine 6 G gain stream (J) is optically pumped by an CW argon laser, the second stream (SA) contains two saturable absorbers, DODCI and malachite green. An acousto-optic deflector (C) dumps single pulse at repetition rate up to 1 MHz. These pulses are compressed through a grating pair (CI) to a duration under one picosecond.

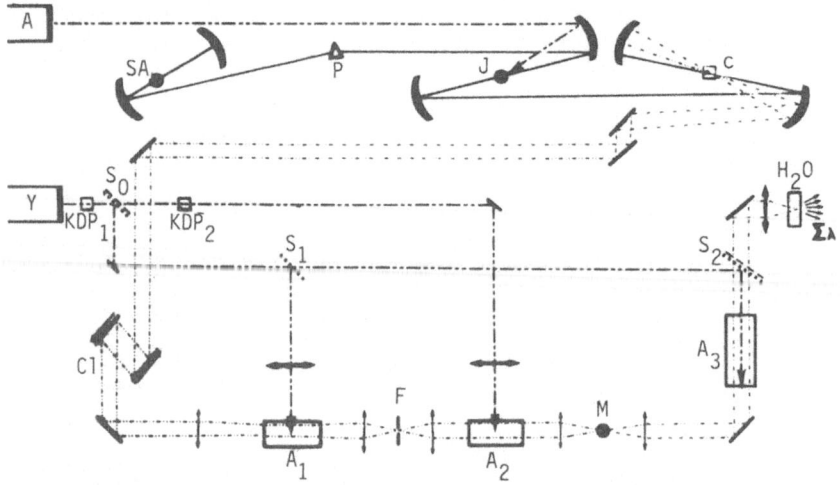

Fig. 1. Experimental arrangement : passively mode-locked CW dye laser and subpicosecond amplifier.

Half of the pulse energy is sent to a real time display of the pulse [5]. In the fig. 2, we present a typical autocorrelation trace, one division representing one picosecond duration. The other part of the pulse is sent to a three stage dye amplifier [6] (A_1, A_2, A_3) pumped by a frequency doubled Q-switched Nd-YAG laser (Y) in synchronism with the dumping of a subpicosecond pulse.

With this arrangement, we currently achieve output energies of about 1 mJ with 0.7 psec pulses at a repetition rate of 10 Hz.

Spectrometric Method : By focusing half the amplified pulses into a 2 cm water cell, we have generated subpicosecond (continuum) pulses with a wavelength content extending from 0.30 μm to 1.2 μm.

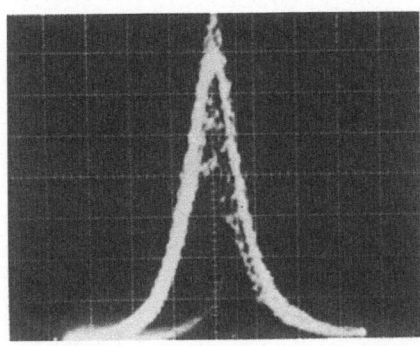

Fig. 2. Real-time autocorrelation:
1 psec/div.

The experimental set-up is shown fig. 3. Half the amplified pulses from the subpicosecond laser (L) is split into a 2 cm cell of H_2O. The residual pulses pass through a variable delay line and are focused to excite the sample (beam 2). The delay is automatically and repetitively scanned by a stepping-motor-driven stage (M). The emerging continuum beam is split into two parts (beams 0 and 1). Both are focused into the sample (E) but only the beam passes through the excited volume. Beams 0 and 1 are then focused on the entrance slit of a 0.25 m spectrograph used in the second order of a 150-groove/mm grating. The resulting two dispersed spectra are recorded by a SPEKTRONIK OSA II Optical multichannel analyser system.

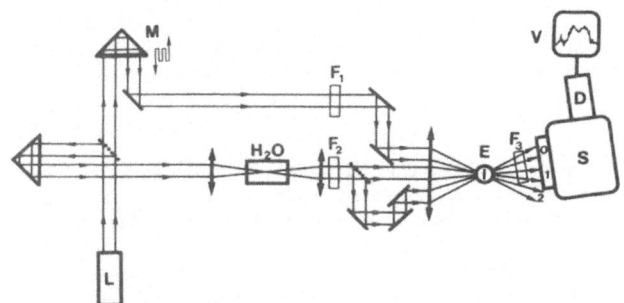

Fig.3.

Spectrometric set-up

3. Results

With this spectrometer adjusted to cover the range 450–550 nm, we measured the transient bleaching induced in a 2 mm thick sample of 500 µM HbCO solution by pulses at a wavelength of 614 nm. For a given delay time, we obtain directly a full double-beam transient absorption spectrum.

The target of the optical multichannel analyser (500 x 400 photodiodes) is electronically divided into two (500 x 200 photodiodes) tracks.

In these conditions, the spectral resolution is 1 nm. The temporal resolution of our system has been checked by measuring the rise of the gain (fig. 4) in a solution of Nile Blue excited by the subpicosecond optical pulse at 613 nm. The medium response can be considered as instantaneous because corresponding to the depletion of the first excited electronic

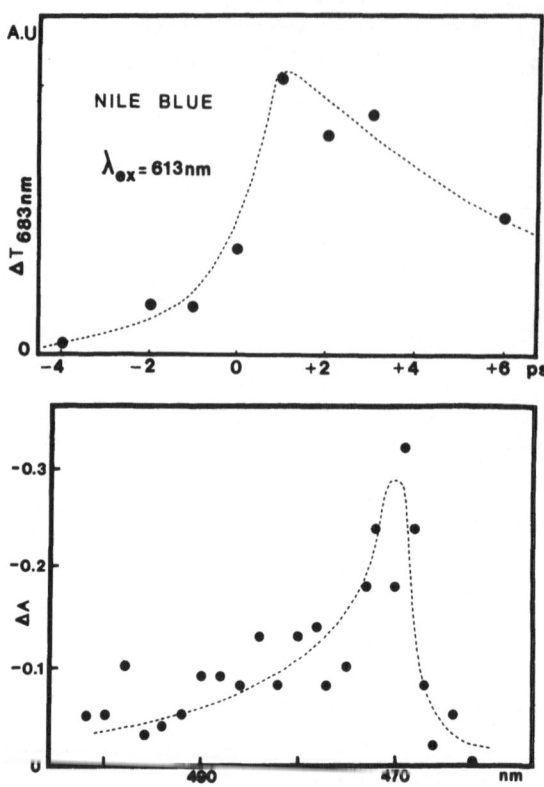

Fig. 4.

Rise of the gain in Nile Blue

Fig.5 . Spectra of absor-
bance changes in HbCO sample 10 psec after excitation

state. As a consequence this curve follows the integral of the temporal correlation of the excitation at 613 nm and the continuum around 683 nm ; its
rise gives the ultimate accuracy which can be reached with our arrangement.

Figure 5 shows the change of absorbance in the 450-550 nm region after
pumping the HbCO sample with the 613 nm pulse 0.7 psec in duration. The two
200-channel tracks of spectral information are digitally processed in the
following way : dark current spectra are subtracted from the raw data ; the
excited track (1) is then normalized by the reference track; the result is divided by a corresponding ratio spectrum obtained with no excitation pulse ;
the logarithm of the new ratio spectrum is taken. This gives the negative
change in absorbance in the excited sample volume as a function of wavelength,
10 psec after the excitation. The build-up of the ΔA spectrum occurs in a
time shorter than 10 psec.

4. Discussion

The reported quantum yield for the dissociation of HbCO is about 0.5 [7] ,
this means that HbCO easily dissociates into Hb and CO. Under our excitation
conditions of 0.7 psec pulses and high intensity ($\sim$ 50 GW/cm^2), the spectrum
shows that Hb-like species are formed in less than 10 psec after irradiation.
This result is in agreement with the data obtained by GREENE et al [2] while
the experimental conditions are different.

Moreover we did not find significant differences between the changes of absorbance at 10 psec and at 1 psec after the pulse excitation (curve not shown).

This seems to show that the dissociation time is very brief as reported by SHANK et al [1] . Neverthless, it will be very hazardous to interpret this result in words of displacement of the iron atom and of the nearby histidine complex. In fact, the transient spectra we observed immediately after excitation correspond to Hb-like species ; that means the light intensity of excitation light induced structural changes of hemoglobin. This may be in agreement with the interpretation of GREENE [2] (who observed a broadening of the Soret and visible bands to about twice their normal width in the 0-680 psec after excitation), while we did not observe such broadened bands in our experiments. Obviously, these results are preliminary data and further studies should be done to elucidate these conflicting points.

Acknowledgments

We thank Prof. Vigneron who supplied us with hemoglobin samples.

References

1. C.V. Shank, E.P. Ippen and R. Bersohn, Science, 193, 50 (1976).
2. B.I. Greene, R.M. Hochstrasser, R.B. Weisman and W.A. Eaton, Proc. Natl. Acad. Sci. 75, 5255 (1978).
3. M.F. Perutz, J. Crystal Growth 2, 54 (1968).
4. C.V. Shank, E.P. Ippen, Appl. Phys. Lett. 24, 373 (1974).
5. R.L. Fork and F.A. Beisser, Appl. Opt. 17, 3534 (1978).
6. E.P. Ippen and C.V. Shank in "Picosecond Phenomena", Springer Series in Chemical Physical, Vol. 4, Ed. C.V. Shank, E.P. Ippen, S.L. Shapiro, Springer Verlag, Berlin, (1978).
7. W.A. Saffran and Q.H. Gibson, J. Biol. Chem. 252, 7955 (1977).

Index of Contributors

Dye Lasers

Editor: F. P. Schäfer
2nd revised edition. 1977. 114 figures. XI, 299 pages
(Topics in Applied Physics, Volume 1)
ISBN 3-540-08470-3

Contents: F. P. Schäfer: Principles of Dye Laser Operation. – B. B. Snavely: Continuous-Wave Dye Lasers. – C. V. Shank, E. P. Ippen: Mode-Locking of Dye Lasers. – K. H. Drexhage: Structure and Properties of Laser Dyes. – T. W. Hänsch: Applications of Dye Lasers. – F. P. Schäfer: Progress in Dye Lasers: September 1973 till March 1977.

Picosecond Phenomena

Proceedings of the First International Conference on Picosecond Phenomena, Hilton Head, South Carolina, USA, May 24–26, 1978
Editors: C. V. Shank, E. P. Ippen, S. L. Shapiro
1978. 222 figures, 10 tables. XII, 359 pages
(Springer Series in Chemical Physics, Volume 4)
ISBN 3-540-09054-1

Contents: Interactions in Liquids and Molecules. – Poster Session. – Sources and Techniques. – Biological Processes. – Poster Session. – Coherent Techniques and Molecules. – Solids. – High-Power Lasers and Plasmas. – Postdeadline Papers.

Ultrashort Light Pulses

Picosecond Techniques and Applications
Editor: S. L. Shapiro
1977. 173 figures. XI, 389 pages
(Topics in Applied Physics, Volume 18)
ISBN 3-540-08103-8

Contents: S. L. Shapiro: Introduction – A Historical Overview. – D. S. Bradley: Methods of Generation. – E. P. Ippen, C. V. Shank: Techniques for Measurement. – D. H. Auston: Picosecond Nonlinear Optics. – D. v. d. Linde: Picosecond Interactions in Liquids and Solids. – K. B. Eisenthal: Picosecond Relaxation Processes in Chemistry. – A. J. Campillo, S. L. Shapiro: Picosecond Relaxation Measurements in Biology.

Springer-Verlag
Berlin
Heidelberg
New York

Springer Series in Optical Sciences

Editor: D. L. MacAdam

A Selection

Volume 2
R. Beck, W. Englisch, K. Gürs

Table of Laser Lines in Gases and Vapors

2nd revised and enlarged edition. 1978.
IX, 202 pages
ISBN 3-540-08603-X

From a review of the first edition:
"... Written directly from a computer memory, this book tabulates 4347 laser lines from 0.12 to 1200 μ (a mean 0.3% separation!) by source molecules and gives the approximate excitation conditions for each set of these, and the corresponding reference. For each of these the exact wavelength and the origin of the transition, where available, is given. Intensities are indicated only by the mark for the strongest lines. Altogether, this volume is a Godsend for researchers having this highly specialized type of problem."
Applied Spectroscopy

Volume 8

Frontiers in Visual Science

Proceedings of the University of Houston College of Optometry Dedication Symposium, Houston, Texas, USA, March, 1977
Editors: S. J. Cool, E. L. Smith
1978. 533 figures, 28 tables. XIV, 798 pages
ISBN 3-540-09185-8

These papers cover the "state of the art" in all areas of visual system investigation. The eye is thoroughly considered from a variety of aspects: the cornea, contact lenses, examination of crystalline lens function, ocular pathologies and retinal function. Much of the material deals with the process of vision after information has been coded in the eye. Psycho-physical visual studies are compared to and contrasted with neurophysiological studies of visual functions. A comprehensive section on the development of visual system function completes the presentation. These papers offer absorbing and useful information for a wide range of teachers, researchers and clinicians.

Volume 18

Holography in Medicine and Biology

Proceedings of the International Workshop, Münster, Fed. Rep. of Germany, March 14-15, 1979
Editor: G. v. Bally
1979. 224 figures, 2 tables. IX, 269 pages
ISBN 3-540-09793-7

This volume presents the proceedings of the International Workshop on Holography in Medicine and Biology, held in Münster, Federal Republic of Germany, on March 14-15, 1979. Special emphasis is placed on a comprehensive survey of biomedical applications of holography. While tutorial review papers give an introduction to each main session, special contributions demonstrate the present state of holographic applications in different fields of medicine and biology. In addition, related techniques, such as Moiré-topography, are included for comparison.

Volume 21

Laser Spectroscopy IV

Proceedings of the Fourth International Conference Rottach-Egern, Fed. Rep. of Germany, June 11-15, 1979
Editors: H. Walther, K. W. Rothe
1979. 411 figures, 19 tables. XIII, 652 pages
ISBN 3-540-09766-X

This book contains the invited papers and postdeadline contributions to the Fourth International Conference on Laser Spectroscopy (FICOLS). It covers original research done by renowned scientists into the application of lasers to spectroscopic problems and related areas. The subjects discussed include

- Fundamental physical applications of laser spectroscopy
- High-resolution spectroscopy, new techniques with sub-Doppler resolution
- Coherent transient and time domain spectroscopy
- Highly excited states, multistep processes in atoms and molecules
- High intensity interactions, laser induced collisions
- Resonant parametric mixing processes
- New laser sources for spectroscopy

Springer-Verlag
Berlin
Heidelberg
New York